# INTRODUCTION TO COMPUTER ARCHITECTURE AND ORGANIZATION

# INTRODUCTION TO COMPUTER ARCHITECTURE AND ORGANIZATION

**HAROLD LORIN**
**IBM Systems Research Institute, New York**

**A Wiley-Interscience Publication**
**JOHN WILEY & SONS**
New York · Chichester · Brisbane · Toronto · Singapore

**Library of Congress Cataloging in Publication Data:**

Lorin, Harold.
  Introduction to computer architecture and organization.

  "A Wiley-Interscience publication."
  Includes bibliographies and index.
  1.  Computer architecture.    I.  Title.
QA76.9.A73L67 1982        621.3819′52        82-8640
ISBN 0-471-86679-2                AACR2

Printed in the United States of America

10  9  8  7  6  5  4  3  2

# PREFACE

This book is an introduction to computers for people who know something about computing. It is aimed at the population of computer professionals and paraprofessionals, casual users, and all others who have some view of computing but know nothing about the logical and organizational nature of a computing device.

As the computer field has matured, it, like all advancing areas of knowledge, has fragmented into a number of specialties. There are a number of very competent computer professionals who know many aspects of software specification, data base design, large application development technique, and so on, but do not know much about the instrument on which their miracles are performed. There is also a large population of computing professionals who know about the technology of computer manufacture or packaging digital technology but have never formed a complete image of the nature of the device they manufacture or whose subassemblies they design. Finally, there is the growing population of casual users whose curiosity has been piqued by their contact with computers and who would like to know more about them.

This book is for all these people. It assumes that the reader has some knowledge of languages like BASIC, FORTRAN, and COBOL, and, starting with known concepts, guides the reader through various stages of computer hardware architecture and organization. It is my opinion, however, that the slightest familiarity with the general ideas of higher-level languages is a sufficient starting point.

This book was developed over a period of years as a result of my interaction with both graduate and professional students. A number of these students reviewed the manuscript, and the extent of coverage and many specific details are the result of their comments and class interaction.

What I try to do here is discuss ideas of architecture and organization generally to provide an overview of several immortal concepts. While specific architectures and machine organizations are mentioned in the book, for example, the IBM S/370, IBM Series/1, IBM 8100, Sperry Rand UNIVAC 1100, Cray Research CRAY-1, Digital Equipment Corporation VAX-11, Data General NOVA, INTEL 8080, and INTEL 8086, the book is not intended to describe specific machines but to use them as examples of architectural and organizational concepts. The danger in using specific examples is that the concepts they embody seem to age with the machines though the concepts are in fact ageless.

There are two great challenges in undertaking a book of this type. One, of course, is content. Naturally I hope that the content is well chosen, the omissions not disqualifying, and the simplifications appropriate. Aside from content, the greatest challenge is sequence. Topics are so interrelated that it is difficult to discuss many points without anticipating later ones. Where necessary, brief characterizing introductions are offered in one area to support discussion in another. This leads to some repetition, but the idea is to give more and more detail each time a topic is discussed. The sequence of chapters was discovered dynamically in class; it has been changed many times, and I am now confident that the conceptual flow of the book minimizes the temptation for readers to look ahead.

HAROLD LORIN

*New York*
*September 1982*

# CONTENTS

PART ONE   ARCHITECTURE

1. **Architecture, Organization, and Implementation**       **3**

    1.   Architecture, 3
    2.   Elements of an Architecture, 5

        2.1.   Mode of Data Representation, 5
        2.2.   Size of the Basic Data Structure, 5
        2.3.   Addressing Conventions, 6
        2.4.   Register Model, 6
        2.5.   Instruction Set, 7
        2.6.   Interrupt Mechanism, 9
        2.7.   Control States, 9
        2.8.   Input/Output, 10

    3.   Organization, 10
    4.   Implementation, 11
    5.   Architecture versus Organization, 11
    6.   Languages and Language Processing, 12
    7.   Hardware/Software Interfaces, 13

2. **Compilation**       **17**

    1.   Introduction, 17
    2.   Initial Statement Processing, 17
    3.   Field Name Recognition, 18
    4.   Symbol Table and Operand Stack, 18
    5.   Statement Type, 19
    6.   Right-Hand Processing, 20
    7.   Intermediate Form, 20
    8.   Next Statement, 21
    9.   Code Generation, 21
    10.   Post-Compilation Processes, 25
    11.   Final Comment, 25

**3.   Data Coding and Reference**                                          **27**

   1.   Data Declaration, 27
   2.   Data Modes and Representation, 27

        2.1.   Pure Binary Characteristics, 29
        2.2.   Binary-Coding Schemes, 30
        2.3.   Floating Point, 30

   3.   Addressable Units, 31

        3.1.   Basic Word Size, 32
        3.2.   Address Definition, 33
        3.3.   Word Sizes and Processor Type, 34

   4.   Nonscalar Representation, 34
   5.   Self-Describing Data, 35

**4.   Register Organization**                                              **37**

   1.   Registers, 37
   2.   Register and Register Model Characteristics, 38
   3.   Register Operand Referencing, 39
   4.   Operations on Registers, 41
   5.   Multiple Register Sets, 42
   6.   Operand Stacks, 43
   7.   Stacks and Compilation, 49
   8.   Some Example Operand Register Models, 52

        8.1.   CRAY-1, 52
        8.2.   UNIVAC 1100, 53
        8.3.   VAX-11, 54

   9.   Software Register Conventions, 54
  10.   Basic Control Registers, 54
  11.   Placement of Control Information, 55

**5.   Memory Addressing Conventions**                                      **57**

   1.   Addressing Conventions in Instructions, 57
   2.   Fundamental Issue, 57
   3.   Address Size Compression, 58
   4.   Fewer Addresses, 60
   5.   Mixed Addressing Forms, 61
   6.   Registers in Address Formation, 62
   7.   Indexing, 63
   8.   Page, Base, and Segment Addressing, 66
   9.   Segmentation, 68
  10.   Indirect Addressing, 69
  11.   Memory Protection, 70

11.1.   Protection Keys, 71
11.2.   Protection Rings, 71

**6. Instruction Sets                                                                   74**

1. **Instruction Set Organization, 74**
2. **Informal Groupings of Instructions, 74**
3. **Other Instruction Groupings, 79**
4. **Symmetric Instruction Sets, 79**
5. **Example Instruction Forms, 81**

5.1.   UNIVAC 1100, 81
5.2.   Data General NOVA, 82
5.3.   Other Formats, 83

**7. Changes in Program Sequencing                                    86**

1. **Sequence Alteration, 86**
2. **Branching, 86**
3. **Condition Code Logic, 89**
4. **Skip Logic, 92**

**8. Subroutine Linkage                                                          95**

1. **Concept of a Subroutine, 95**
2. **Subroutine Linkage, 96**
3. **Basic Linkage Instruction, 96**
4. **Basic Stack Linkage Control, 99**
5. **Parameter Passing, 101**
6. **Activation Record, 102**
7. **Stacks, Parameters, and Activation Records, 103**
8. **Addressing and Linkage, 105**

**9. Interrupt Mechanisms and Control States                      107**

1. **Interrupt Mechanisms, 107**

1.1.   Introduction to Interrupt Response, 107
1.2.   Interrupt Inhibition, 109

2. **Variations in Interrupt Architecture, 109**
3. **General Description of S/370 Interrupt Architecture, 110**

3.1.   Interrupt Classes, 110
3.2.   The S/370 Program Status Word, 111
3.3.   Organization of Lower Memory, 113
3.4.   Interrupt Response, 114

4. **Other Interrupt Schemes, 115**
5. **Control States, 116**

**10.  Virtual Memory**                                                **119**

    **1.  Concepts of Virtual Memory, 119**
    **2.  Linear Virtual Memory, 121**

        2.1.  Basic Mechanisms, 122
        2.2.  Memory Space Management, 123
        2.3.  Address Formation in Paging Systems, 124
        2.4.  Page Faults, 125
        2.5.  Linear Virtual Memory and Protection, 127
        2.6.  Segmentation in Linear Virtual Memories, 127
        2.7.  Associative Memory Assist, 127

    **3.  Alternative Architectural Concepts, 129**

**11.  Input/Output**                                                  **132**

    **1.  Input/Output, 132**
    **2.  Disk Characteristics, 134**
    **3.  Input/Output Architecture Features, 136**
    **4.  Basic Processor I/O Architecture, 137**
    **5.  Operating Systems Services, 140**
    **6.  Input/Output Supervisor Flow, 142**
    **7.  Control Languages, 143**
    **8.  Higher Levels of Input/Output Support, 145**
    **Bibliography, 146**

**12.  The Power of an Architecture**                                   **147**

    **1.  Concept of Architectural Power, 147**
    **2.  The Army-Navy Study, 148**

        2.1.  Absolute Criteria, 149
        2.2.  Quantitative Criteria, 151

    **3.  Measures of an Architecture, 154**
    **4.  Functional Instructions, 155**
    **5.  Architectural Comparisons, 156**
    **6.  Ideal Architecture, 158**

        6.1.  Addressing and Memory Referencing, 159
        6.2.  Characteristics of Ideal Architecture, 159

    **7.  Architecture, Compilability, and Design Complexity, 161**
    **8.  The Great Debate, 163**
    **9.  Level of an Architecture, 163**

        9.1.  Support of an Operating System, 165
        9.2.  Extension Toward Higher-Level Languages, 166
    **Bibliography, 168**

## PART TWO   ORGANIZATION AND IMPLEMENTATION

**13.   Concepts in Organization and Implementation**                          173

   1.   Introduction, 173
   2.   Organization and Implementation Decisions, 174

      2.1.   Width of Data Paths, 174
      2.2.   Degree of Circuit Sharing, 174
      2.3.   Definition of Specialized Units, 175
      2.4.   Parallelism of Functional Units, 175
      2.5.   Buffering and Queuing, 175
      2.6.   Prediction, 175
      2.7.   Underlying Technology, 176
      2.8.   Functional Implementation, 176

   3.   Price/Performance Goals, 176
   4.   Architecture, Organization, and Price/Performance, 177

**14.   Basic Concepts of Instruction Execution**                              180

   1.   Basic Cycles, 180
   2.   Execution of an Instruction—Instruction Stages, 182
   3.   Instruction Stages and Machine Cycles, 186

**15.   Organization for Increased Performance**                               189

   1.   Organizational Approaches to Faster Machines, 189
   2.   Instruction Times, 190
   3.   Cycles Reduction by Stage Speedup, 190
   4.   Stage Redefinition, 191
   5.   Resequencing Decode and Address Formation, 192
   6.   Concluding Remarks, 195

**16.   Extended Lookahead**                                                   197

   1.   Inter-Instruction Overlap, 197
   2.   Storing Values into Memory, 202
   3.   Memory Read and Write Contention, 204
   4.   Memory Reference Delays, 204
   5.   Increasing Lookahead, 206
   6.   Details of I/E Function and Relationship, 207
   7.   Buffering and Extended Lookahead, 210

**17.   Parallel Instruction Execution**                                       214

   1.   Multiple Instruction Execution, 214
   2.   Considerations in Multiple E-Box Design, 215
   3.   Populations of E-Boxes, 216

**4.    Delivery of Instructions to E-Boxes, 217**
**5.    Delivery of Operands to E-Boxes, 218**
**6.    Relationship between E-Boxes and Registers, 219**
**7.    Instruction Sequencing, 220**

    7.1.    Sequencing Techniques, 221
    7.2.    Intersecting Memory References, 226

**8.    Code Equivalence and Rearrangement, 227**
**9.    Characteristics of E-Unit Designs, 228**

    9.1.    Single-Level Units, 228
    9.2.    Multilevel Pipeline Units, 229
**10.    Vector Manipulation, 231**

**11.    Multiple I-Stream Machines, 233**

**18.    Memory Organization                                       238**

**1.    Basic Goals of Memory Design, 238**
**2.    Enough Memory, 238**
**3.    Memory Organization, 239**

    3.1.    Banking, 240
    3.2.    Interleaving to Speed Memory Response, 241

**4.    Memory Lookahead, 242**
**5.    Memory Partitioning Techniques and Processor Lookahead, 243**
**6.    Limits on Memory Partitioning, 243**
**7.    Memory Times, 244**
**8.    A Memory/Processor Interconnect Organization, 245**
**9.    Alternative Interconnection Organization, 248**

**19.    Memory Hierarchies                                        251**

**1.    Notion of Hierarchy, 251**
**2.    Instruction Buffers, 253**

    2.1.    Basic Buffer Fill Technique, 256
    2.2.    Flexible Priority Driven Buffer Fill, 257
    2.3.    Short Loop Mode, 257

**3.    Branch Instructions and Instruction Buffering, 257**
**4.    Operand Buffers, 259**
**5.    Cache, 260**

    5.1.    Cache Mapping and Loading, 262
    5.2.    Cache Contents Management, 265
    5.3.    Stores and Changes to Cache, 266
    5.4.    Some Concluding Comments, 266

    6.  Other Hierarchical Notions, 267
    7.  Instruction and Control Memories, 268
    8.  Final Comments on Memory, 269

## 20.  Input/Output Design          271

    1.  Basic Concepts, 271
    2.  Programmed Input/Output, 272
    3.  Processor Overlap, 276
    4.  More Elaborate Input/Output Design, 278
    5.  Buffered Input/Output, 281
    6.  Direct Memory Access Organizations, 281
    7.  Multiple Units, 282
    8.  General-Purpose Channels, 284
    9.  Input/Output Processor Concepts, 286
  10.  Processor Support of Input/Output Functions, 288

## 21.  Overview of Implementation          290

    1.  Introductory Observations on Technology, 2

        1.1.  Physical Change, 290
        1.2.  Performance and Price/Performanc    92
        1.3.  Storage, 292
        1.4.  Summary Remarks, 294

    2.  Computer Building Blocks, 294
    3.  Architecture, Design, and Implementation,
    4.  Microprocessors and Microcomputers, 297
    5.  Basics of Implementing One Architecture w      ther, 299
    6.  Considerations in Implementing One Archi    ure
        with Another, 301
    7.  Microcode Types, 303
    8.  Extending an Architecture, 304
Bibliography, 305

**INDEX**          307

# PART ONE

# ARCHITECTURE

Chapter 1

# Architecture, Organization, and Implementation

## 1. ARCHITECTURE

The exact meaning of the word *architecture* in the context of computers is a little uncertain. In general, architecture refers to the visible characteristics of a system as seen by a person or a program creating code capable of running on the machine. It is common, however, to use the word to mean the view of a machine shown by its *assembly language*. An assembly language is a low-level programming language in which the basic characteristics of a computer system are more directly represented than in a language like COBOL or FORTRAN. The assembly language programmer is aware of memory locations used in the machine, the actual instructions of a machine, and possibly the general speed of instructions. He* is not necessarily aware of the underlying organization of the machine in terms of operational logical units or of the hardware technology used in the construction of the machine.

The word architecture is frequently used to mean the visible characteristics of only the element actually performing instructions, that is, the *processor* of a computer system. Since a programmer may view other elements of a computer system through the processor, architecture also more generally means the characteristics of all the component elements of a system that might concern a programmer working at machine level. These components include processor(s), *memory*, and *input/output subsystems*.

A processor is a unit that interprets instructions and changes data in conformity with the instructions of a program. A *computer* may have one or more processors either dedicated to the execution of specialized instructions or capable of performing all instructions. Memory is a device that holds instructions to be executed and data to be operated on. Input/output subsystems are collections of units that connect processors and memories with the devices (1) that interface with the outside world (printers, terminals, sensors, etc.) or (2) on

---

* Throughout this book "he" is used to mean "he" or "she." This avoids the cumbersome "he/she."

which large amounts of additional data or instructions are stored and from or to which data is moved to or from memory.

The word architecture is also used to suggest the relationships among various building block elements of hardware and software systems. A system may be thought of as having several *architectural levels*. Figure 1 shows a representation of architectural levels. Each rectangle represents a set of functions and the horizontal lines are the interfaces between the functions. The first part of this book discusses the architectural level represented by the line labeled *machine-architecture interface*. The second part of the book addresses the levels that support this interface.

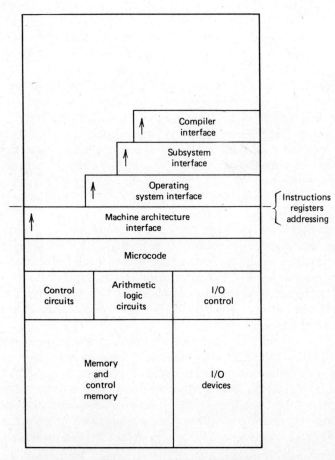

**Figure 1.** Architectural levels.

## 2. ELEMENTS OF AN ARCHITECTURE

From the point of view of a programmer looking at a computer system from an architectural level, a system has a set of characteristics that define how data will be represented and referenced, the operations that can be performed on data, and other features that determine how sequences of operations must be executed to achieve a computational result.

### 2.1. Mode of Data Representation

This refers to the representation of values in the processor and memory and the manner in which strings of binary digits should be interpreted at different points in the architecture. The representation of information may be in either pure base 2 (binary) or encoded form generically called the *binary-coded decimal*. There is also a form of scientific notation, called *floating point*, where values and exponents may be represented separately in either pure binary or coded form.

*examples?*

When a value is to be manipulated by a processor, the circuits of the processor will be designed to treat the bits of that value as either pure binary, binary-coded decimal, or floating point. For example, a unit designed to add binary numbers will produce an erroneous result for the logic of a program if the data presented to it is in binary-coded decimal form. As another example, a program must know the mode of representation for various values printed on a system printer. Devices like printers always require data to be in some coded decimal form, which may not be the form in which the data is arithmetically manipulated in the processor. The program must issue instructions to put the data in proper form for printing.

### 2.2. Size of the Basic Data Structure

This refers to the organization of values in the memory of the system and in the processor. The basic architectural data structures are *bits, bytes, characters, and words*.

A computer system's memory is organized into *locations*, which may be visualized as elements in a one-dimensional array. Each element has a name, or an *address*. Thus we speak of a memory having 4096 locations; each location can be directly referenced according to its position on the list. For example, there are memory locations 0000, 0001, 0002, . . . , 4094, 4095. When one of these numbers appears in control circuits connected to memory either the contents (the value that exists in the location) will be brought from memory to the circuits of the processing unit or information in the processing unit will be stored in the memory, depending on the work associated with the address.

Each location in a memory has a fixed size that represents the number of *bits* (binary digits) held in that location. Thus we speak of machines with 8, 16, 24, and 32-bit memories. This describes the number of bits that will be transferred between the memory and the processor at a single memory location as the result of an instruction referring to that memory location.

## 2.3. Addressing Conventions

An address represents a particular location in memory. In most architectures the address of an instruction or an operand in memory may be formed in various ways. The most basic representation of an address is in the part of an instruction called the *address field*. However, there are various means for modifying an address field by values in other parts of the processor, for using complete addresses not contained in an instruction, and so on. A key feature of an architecture is the number of ways *address arithmetic* may be used to form addresses to memory. These addressing conventions involve the use of values held in various parts of the processor that may be concatenated, added, or subtracted with the contents of the address field in the instruction. The addressing schemes are used to provide relocatability of programs (the ability to move them around in memory), to provide protection (the ability to constrain programs from referring to certain memory locations), and to reduce the space required to represent an address in an instruction.

## 2.4. Register Model

This refers to the number, type, size, means of addressing, and relationships within a set of storage areas that form specialized working spaces distinct from main memory. The specification of the *register model* of an architecture is a basic architectural feature. An architectural function of registers is to provide a working space that can be referenced by an instruction in a highly compressed way, reducing the number of bits necessary to refer to an operand. An operand for an arithmetic operation can be referenced in fewer bits if it is in a register than if it is in memory.

Registers are of two general types—*operand* and *control*. An operand register is used to hold operands that are to be used during the computational flow of a program. A control register contains addresses or other values that represent the status of the machine as distinct from the status of data values used by the machine.

Registers are used for many purposes, some of the most common types of which are:

*Arithmetic Registers (Operand Registers).*   Working areas where data involved in computational operations or data that is being moved from one location to another may be temporarily stored. Operand registers are never conceptually visible to a programmer in a high-level language. Their use is fundamental to machine level or assembly language programming.

*Index Registers.*   These registers are used to modify an address value that exists in an instruction by adding or subtracting a value to form a final memory address. Index registers are commonly used to support the notion of a subscript value such as A(I) or an index value such as DO I = 1, 10, 1. If the array A has 10 elements, the first element may be in memory location 1000, the second element in 1001, and so on. An index register will provide a means of consecutively addressing A(1), A(2), and so on without modifying the operand address of instructions operating on elements of the array.

*Control Registers.* This is a generic name for a class of registers used to represent the status of the machine. A register holding the address of the next instruction to be executed, the *instruction (control) counter*, is a control register. Various control registers are used in connection with program flow, addressing, reporting machine status, and so on. We will discuss these primarily in the context of the control functions they perform.

## 2.5. Instruction Set

This is the collection of specific *operations*, such as ADD, MOVE, BRANCH, STORE, and LOAD, that, in sequence, form a machine program and define the pattern of data movement and transformation. An architecture has a unique set of operations and conventions for determining the locations of data on which the operation is to be performed, which are represented in *instructions*.

Various architectures differ in the specific operations they provide and in the methods they use to refer to data that will be manipulated by an operation. In general, an instruction has the form OPERATION, OPERANDS. OPERATION specifies the function that is to be performed; OPERANDS provide a way of designating the current location of data the operation is to be performed on.

Instructions on a computer are executed in a sequence determined by their location in the memory. In most computers (those that are members of a class of computer called Von Neuman machines), instructions and data are intermixed in memory locations. A control register in the processor, sometimes called the *instruction counter* or *program address counter*, contains the location of the next instruction to be executed. When an instruction execution sequence begins, the instruction whose address is in the instruction counter is brought from memory to a storage area called an *instruction register*. The instruction is interpreted by *decode circuits* that provide for electronic signals to be developed in the processor as a result of the value of the operation field. These signals, or sequences of signals, result in the execution of the instruction. Execution is the application of the function of the operator on the operands. When the execution of an instruction is complete, the instruction counter is advanced to the address of the next instruction. This instruction is then brought from memory to the instruction register and executed.

Instruction execution sequence may change as a result of an instruction that directs a *branch* (also called *transfer* or *jump*). These instructions contain the address of the next instruction to be executed rather than the address of an operand. They provide for changes in program flow as a result of conditions in the data. The high-level programming concept of IF (test for a stated condition and alter program flow if it obtains) universally translates into some type of branching instruction.

The following illustrates the execution of instructions on a machine that has a set of operand registers and a set of index registers, and that must bring one operand to a register to perform arithmetic. On such a processor, the statement A = B + C will lead to the compiler generation of the coding shown below. (Assume MEM1 is symbolic representation of the memory address holding B when the program is executed, MEM2 holds C, and MEM3 will hold the result, A. X1 shows that addresses will be modified by the contents of index register 1.)

| Op Location | Op | Register | Operand Loc (Symbolic) |
|---|---|---|---|
| 0100 | LOAD | R1 | MEM1 (X1) |
| 0101 | ADD | R1 | MEM2 (X1) |
| 0102 | STORE | R1 | MEM3 (X1) |

This is the assembly language form where the operation code (e.g., LOAD) and addresses are in symbolic form. The actual contents of a machine location holding an instruction will have a specific format that might be unlike the format of the assembler. The machine format for the above instructions might look as follows:

| 0–5 Bits: Op Code | 6–9 Bits: Registers Designation | 10–13 Bits: Index Designation | 14–31 Bits: Memory Address |
|---|---|---|---|
| 011101 | 0001 | 0001 | 000000011111010000 |
| ADD | R1 | X1 | 1000 |

The sequence of events caused by the three instructions is as follows: The instruction counter initially reads 0100 (decimal) at the start of this sequence. The LOAD is brought to the processor and interpreted. Because of the designation of X1 as an index register, the contents of X1 is added to MEM1 in order to form the actual indexed memory address. The value at MEM1 + (IX) is placed in register 1. Then the instruction counter is automatically augmented to read 0101 (we do not have to execute an instruction to do it), and the ADD instruction is brought to the processor. The interpretation and execution of the ADD instruction causes the contents of R1 to be added to the contents of MEM2 (as modified by the value in X1) and the result returned to R1. The next instruction, in 0102, causes the result in R1 to be stored in MEM3 (as modified by the value in X1).

The operation codes specify what is to be done. The register designation is necessary because the example architecture has more than one register in which operands may be placed so that the specific register to be used must be designated. The index register designation chooses an index register whose value is added (in this example) to the value in the address field (MEM) to form an address for A, B, and C. Thus the memory locations of A, B, and C are:

| Variable | MEMn | X1 | Final Memory Location |
|----------|------|-----|----------------------|
| A | 1000 | 500 | 1500 |
| B | 1001 | 500 | 1501 |
| C | 1002 | 500 | 1502 |

Because index register 1 is specified in the instruction, its contents of 500 are added to MEM1, MEM2, and MEM3 before a request is made to memory for B and C and before the reference to A. We do not show how the value 500 was placed into index register 1. Some previous instructions for placing values in index registers can be assumed to have been performed.

## 2.6.  Interrupt Mechanism

Many architectures define events that can cause the normal flow of instruction execution to be altered without the use of a branching or jumping instruction. The program gives no indication of a desire to break the flow of instructions.

While a provision for interrupting the flow of instructions because of an event in the system is not required, many computers contain a provision for an immediate response to an external event or to conditions occurring in the machine. Some programming languages present a similar idea with the ON CONDITION statement.

When a designated event, such as an operator pressing a key on a console, a value exceeding a range limit, or the completion of an input/output operation, occurs, a signal causes the next instruction executed to come from a location out of the normal sequence. This location, as designated by the architecture, holds the first instruction (or the address of the first instruction) dealing with the condition. Part of the architecture also defines the ways that the interrupted instruction flow can be resumed without logical error or loss of data.

## 2.7.  Control States

Many architectures define special modes of operation called *control states*. A set of instructions can be defined that will only be executed when the processor is in a control mode. A set of rules for addressing memory may also be developed that restrict the memory locations to which a program not in control state may make reference.

The intent of the feature is to provide for programs to share a machine by coexisting in memory and having alternating periods of time when they are executing. This mode of operation is called *multiprogramming* or *time sharing*. The goal is to provide access to a computer facility for many users at the same time, either to enhance the effective use of the machine or to provide continuously available and economical computer service to a population of users.

In such an environment it is common to have *operating system* programs occupy some of the memory. The function of the operating system is to coordinate the execution of the sharing programs and to prevent one program from damaging another. A control state is defined for use by the operating system programs only. In this way operations that define the basis of sharing can be performed only by the operating system in a control state. Control operations include determining what memory will be available to individual sharing programs, how shared input/output devices will be used, and what time periods will be made available to each program.

### 2.8. Input/Output

*Input/output (I/O) architecture* determines the methods and forms used to transfer data between a processor/memory and the I/O devices of the system. Rules for referencing I/O devices, for defining where data are to be moved from and how much data is to be moved, and for how completions and malfunctions are to be reported are specified in the I/O architecture.

### 3. ORGANIZATION

Computer organization, sometimes called *computer design* (we will use the words organization and design interchangeably) determines the nature of the operational units that realize the architectural specifications. There are choices about the specific sequence of machine events that constitute the execution of an instruction, about the amount of data that can be handled at one time, and about the partitioning of functions among the building block elements of a system.

The designer of a computing system is influenced on one side by the available implementation technology and on the other by the architecture. It is the designer's job to organize the architecture so that it is appropriate for the technology.

For example, it is an architectural decision whether a computer will have a multiply instruction. It is an organizational decision whether that multiply instruction will be performed by a special multiply unit or by a hardware algorithm that makes repeated use of the add unit of the machine. This decision will be influenced by an appreciation for the frequency of multiply operations in programs, the general speed goals of the system, the number of circuits involved in the special multiply unit, the cost of those circuits in current technology, the size of the unit given the number of circuits it must have, and so on.

Another example of organization lies in the selection of the memory speed required for the architecture to perform well. The forms and method of addressing a memory is an architectural choice, but its speed is an organizational choice. The actual technology of the memory is an implementation choice. The architect says, "I wish to be able to address a memory of a certain size"; the designer says, "for a certain price/performance the memory should be of a certain speed"; the implementor says, "I will build this memory from elements with certain electronic and packaging properties."

## 4. IMPLEMENTATION

Implementation refers to the physical concepts used to support the organization. There is a strong relationship between architecture, organization, and implementation, but there are also degrees of freedom so that various architectures can be mapped onto various organizations and implemented in various ways.

Implementation is concerned with selecting the actual electronic and chemical technology to be used. It is also concerned with the physical packaging of circuits and with determining the use of physical space, the number of uniquely designed components, the layout of processors, memories, and data paths on physical *chips, cards,* and *boards.* Other aspects of implementation involve powering, cooling, and methods for manufacturing the components of a system. In this book we are only peripherally interested in implementation.

## 5. ARCHITECTURE VERSUS ORGANIZATION

The notions of architecture, organization, and implementation are three more or less distinct aspects of computer systems. However, there is a variety of opinion about what particular features of a system are architectural, which are organizational, and which are primarily involved in implementation. Some computer systems describe as architectural features what others might consider features of organization or implementation.

One finds architectures that are supported by rather different organizations, organizations that are supported by various technologies, architectures using similar organizations, and so on. Many computer producers market models of the same architecture with different organizations and implementations. Consequently, they have different prices and performance characteristics. One example is the IBM family of systems—all of which have architectures that can be generically termed System/370 architecture. This architecture is offered in various models, not only by IBM but by others as well.

Although there are wide differences in the ways that organization and implementation can support an architecture, there is also an interplay between these elements. For example, an architecture that has good efficiency (determined using measures we will discuss further along) may be too expensive for the organization and implementation methods of a certain technology period. Also, an architecture that is very efficient in a sequential program may restrict the extent to which parallel or overlapped functions can be organized. A less efficient architecture may be preferred because it provides easier organization and can increase the speed of a machine by doing more things in parallel.

In the class of processors called *microprocessors*, (the smallest physical implementations of computer architecture), the relationship between architecture, organization, and implementation is very close. Changes in technology usually result in the development of new and more powerful architectures. This is partially because the users of microprocessors have less of a requirement for generation-to-generation compatibility. In general, at the microprocessor level architecture, organization, and implementation move together and one does not see the generation-to-generation reimplementation of architectures with new organizations and new technology that is common in larger machines.

Organization may be architecturally driven or technologically driven. An architecturally driven organization faces the problem of mapping a given instruction set onto a particular technology. Technologically driven organizations face the problem of how complicated an architecture can be supported by circuits that fit on a single chip.

Sometimes an idea that starts off as an organizational idea, for example, a way of improving performance without changing the visible machine, will be introduced into the architecture at a later time because programmer (or compiler designer) awareness may improve the use of the feature. There is an example of this in the area of small specially managed fast memories whose contents are automatically managed by hardware. To improve the probability that desired data are in the special fast memories, later versions of a machine may have instructions that allow a program to participate in the determination of the contents. When these instructions are put into the machine, the small fast memory becomes part of the architecture.

At any time in the computer industry one will find computer systems that are reimplementations of earlier organizations, ones that are reorganizations of stable architectures, and new architectures. There will always be differences of opinion about whether better, faster, more efficient machines require new architectures, new organizations, or new implementations. There will be differences of opinion about the extent to which compatability with existing systems must be a feature of any new system.

## 6. LANGUAGES AND LANGUAGE PROCESSING

We use the word *software* to mean a set of programs provided with a hardware system to facilitate its use by programmers and operators of the system. This use of the word excludes applications programs from the world of software and includes programs that provide general functions independent of the details of an application. The reader is probably familiar with the fact that there is a wide diversity of opinion in the software world about what applications are and what systems software functions are.

It is the software of a system that largely defines how it looks to an application programmer. As a group, the readers of this book probably view computing through the image presented by a programming language such as BASIC, COBOL, PL/1, or PASCAL. In fact, some systems present no view except through the programming languages. Where certain systems directives that are not formally part of the programming language are required, they are represented in the syntax of the language as consistent extensions of the programming language.

The programs that are written in high-level programming languages must be converted into machine programs. The process of this conversion is called *compilation* and it is accomplished by a software element called a *compiler* or (less commonly) a *language processor*. A compiler that can do this conversion for a particular machine is a precondition for using that language as a programming tool for the machine.

It is usual for a compiler program to reside on an input/output (I/O) device, such as a disk or tape, and be summoned into the processor and memory of a system when a programmer wishes to compile. During the compilation process, the program running on the machine is the compiler. The data input to the compiler is the high-level language program. The output of the compiler is usually a representation of the program in machine executable form or in a form that is very nearly machine executable. Figure 2 shows the stages of compilation from the initiation of compiler processing until the creation of a machine program.

A compiler may not necessarily generate executable machine instructions. Sometimes a compiler will reduce the high-level language to a form that cannot be directly executed. The execution of the output of the compiler depends on the presence of another software component called an *interpreter*. The output of the compiler is loaded into memory and the interpreter program, also in memory, uses the compiler output as input. The interpreter program causes the execution of actual machine instructions that bring about the transformation of data indicated by the *pseudo-instructions* generated by the compiler. The software interpreter program effectively emulates a machine whose architecture is the output format created by the compiler.

## 7. HARDWARE/SOFTWARE INTERFACES

A fundamental concern in the architecture and organization of computers is the *hardware/software interface*. This interface determines how much of the system's function will be implemented using hardware technology and how much using software. The nature of the hardware/software interface determines the ease with which programs to accomplish certain functions can be written and the time it will take for those functions to be accomplished. Instruction sets can be either difficult or easy to compile good code for; operating systems functions can be supported by structures in the architecture either sufficiently or insuffi-

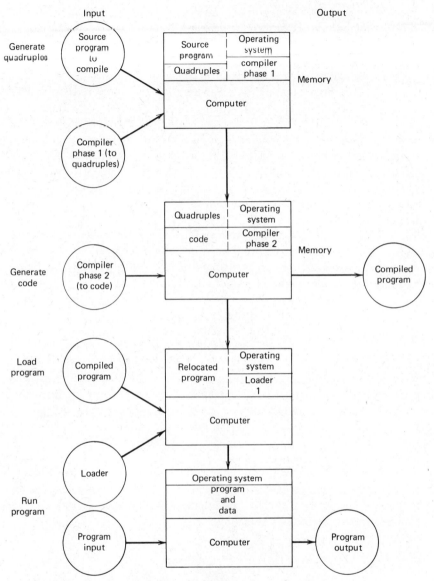

**Figure 2.** Compilation and loading.

ciently; and an adequate mechanism for determining programming errors may or may not be in the hardware architecture.

Computer functions may be implemented in several ways. Machine functions may be hard-wired (fixed by circuit design) or made available through either software or an intermediate technique called *microcode*. Microcode is a method, discussed in the final chapter in this book, that programs the features of an apparent or target machine into the architecture of an underlying machine emulating the target machine. When this is done, the architecture is considered to be the image supported by microcode.

Sometimes microcode provides functions that are not part of the hardware architecture of a target processor but are generally provided in software. For example, portions of an operating system may be put into microcode. A complete compiler or language translator may be provided in microcode. When functions traditionally provided by software packages are provided by microcode, that code is sometimes called *firmware*.

The name suggests that the function is basically delivered through a program but that the program is stored, executed, and otherwise manipulated differently than usual program packages are. The firmware program is written by *microprogramming* the functions in an instruction set that is not visible to the user of the system at the machine architecture level. Firmware programs may be held in a special memory not addressable by a program and encoded in a machine language that differs from the machine language visible to a programmer or compiler. A common way of implementing an interpreter that will process the output of a compiler that does not go all the way to machine code is with an interpreter written in firmware. In some organizations and implementations (e.g., The IBM System/38) the apparent architecture is the set of forms and instructions that a firmware interpreter will accept. The instruction set used to code the firmware interpreter is not visible to a programmer or compiler, and therefore not part of the architecture.

The set of issues at the hardware/software interface ranges from details of how convenient it will be for an instruction to determine why an interrupt occurred to the nature of architectures that can directly execute programs written in high-level languages. If everyone is using COBOL, why have a COBOL compiler? Why not just build a machine that executes COBOL programs? There are many proposals for *high-level language machines* or *language directed architectures* that use combinations of hard-wiring and firmware to extend the architecture of a machine in the direction of forms used by programming languages. There is a similar movement to define programming languages which are more conveniently represented in computer architecture defined by firmware techniques than current languages.

Current arhitectures are almost all influenced by concepts of programming structures found in high-level programming languages or by common functions of operating systems. As more has been learned about programming structures, as hardware has gotten cheaper, and as design and implementation techniques have become richer and more flexible, architectures have tended to be extended under the influence of well-defined programming structures like queues, stacks, and linked lists. However, there is currently a difference of opinion as to what extent architectures should be extended in the direction of programming languages and to what extent operating system considerations should be reflected in architectures. To a certain degree, issues at the hardware/software interface reflect an old argument about compilation vs. interpretation. For many years it has been thought that excellent compilers were necessary and that the compilation procedure was more efficient than the interpretation of a high level language form. Currently, things are not quite so clear and there is renewed interest in extending architectures to close the *semantic gap* between programming and machine architecture concepts.

## QUESTIONS

1. What are the basic elements of a computer architecture?

2. Distinguish between concepts of computer architecture, organization (design), and implementation.

3. Describe how a sequence of instructions is executed on a processor.

4. What is the purpose of an interrupt mechanism?

5. What is the difference between an interpreter and a compiler?

6. What is firmware?

# Chapter 2

# Compilation

## 1. INTRODUCTION

As the reader may suspect, the process of compilation does not lend itself to terse description in the early chapters of a book concentrating on hardware topics. We undertake a very summary description of some of the fundamental features of the process of compilation to provide a conceptual link between high-level languages and basic concepts of hardware architecture. There are basic ideas about how a program that is intelligible to a programmer becomes intelligible to a computer. During this discussion we introduce some ideas about the kind of machine elements that concern us throughout the book. There are vast areas of compiler considerations on which this chapter does not touch.

The processes of compilation are shown with the simplest of programs:

```
BEGIN;
    A = B+C;
    D = A*3;
END;
```

Assume that this program exists on a disk and is to be processed by a compiler. When the operating system causes the compiler to execute, the compiler will read the program from the disk into memory and begin the process of compilation.

## 2. INITIAL STATEMENT PROCESSING

As processing starts, the compiler will recognize the word BEGIN denoting the beginning of a program. It will perform whatever initializing functions it must on its internal data areas to prepare for the compilation of the program.

## 3.   FIELD NAME RECOGNITION

The compiler will recognize the symbol A as the name of a data field. It will recognize this because rules of recognition represented in its own instructions will define a data field name to be the first symbol or set of symbols occurring in a statement that is not a *directive* of the language and which is preceded or succeeded by a *delimiter*.

A directive of the language is a word like IF or DO or BEGIN that suggests a particular programming structure or function. A delimiter is a character defined in the language to be an operator (e.g., +, −. =, *) or a separator (e.g., ( or a space).

Many languages permit names like AMY, that is, names of more than one letter. The compilation program must inspect a string of characters until it discovers a character that cannot be part of a name. In all programming languages the symbols used for operator representation may not be part of a name. In this example, the compiler looks at the character succeeding A, discovers it is an operator symbol, and knows that the full name of the field is A. This process involves comparing each succeeding character to the list of operator and delimiter symbols. When = is recognized, the compiler knows it has collected the complete name of the field A.

When the compiler program recognizes A as a field, it will search a *symbol table* to determine if A has previously been used in the program. If, as in this case, A has not been used, then A must be added to the symbol table.

## 4.   SYMBOL TABLE AND OPERAND STACK

A symbol table contains various pieces of information about a variable. It contains its name, how the bit patterns of its values are to be interpreted (binary, encoded decimal, etc.), and its length in characters or words. Symbol tables may hold a great deal more information about a variable, depending on the sophistication of the compiler and the richness of the architecture for which the program is being compiled.

The symbol table after the compiler has placed A in a position is shown below. Notice that the compiler has been programmed to assume that A is a binary number when it is not told differently. This is the compiler *default*, an assumption that a compiler makes about data unless it is told differently.

Let's assume, for this example, that all scalar variables in this program are going to occupy one memory location in the machine memory.

*Symbol Table*

| Symbol Table Address | Field Name | Size | Mode | Relative Address |
|---|---|---|---|---|
| 001 | A | 1 | Binary | Unassigned |

The placement of A on the symbol table provides a quantity called the *symbol table address* for A. This symbol table address is the place in the symbol table where a description of A may be found. This symbol table address for A, 001, will be used to refer to A throughout the compilation.

We place the symbol table address of A on a list called the *operand stack*, which is discussed more fully later. The operand stack is a last in/first out (LIFO) list where the symbol table addresses of operands are collected.

## 5.  STATEMENT TYPE

The recognition of A as the first element of the statement enables the compiler to define the type of statement it is processing. A statement that begins with a field name may be an *assignment statement*, which must have a certain form. The instructions to recognize the form of the statement are coded into the compiler or represented by certain table values that define language syntax.

Since it knows it is dealing with an assignment statement, the compiler knows that an operator must follow A. It determines that = is a valid and appropriate operator, valid in that it is an operator, appropriate in that the equal function is significant at this point in an assignment statement. If an invalid or inappropriate symbol is found, the compiler develops an error message. In some compilers this will cause an end to their attempts to compile. Other compilers try to forgive some errors or run to the end of the compiling process anyway so that the programmer may see as many errors as possible after one compilation attempt.

Errors of form, where the compiler finds the rules of the language violated, are *syntactical* errors. The program as written is not in accordance with the rules of the language. As the reader has surely experienced, a program may be syntactically correct, conform to rules of language, and still contain semantic or logical errors that are impossible for the compiler to recognize. Among the arguments presented by those who wish to extend architectures is the provision of machine facilities during the running of a program that will detect errors the compiler cannot detect. Among these, for example, are efficient ways of recognizing that the bounds of an array have been exceeded by an array reference and that a reference to an undefined variable (one that has no assigned value) has been made.

The recognition of the = causes the compiler to place it on a list of operators. This list is similar to the operand list on which the symbol table address of A was placed and is called an *operator stack*. An entry on an operator stack contains the operator symbol or the address of a routine that will process the operator symbol later in the compilation process.

At the end of processing the *left-hand side* of the assignment statement, the operand and operator stacks look like this:

| Operand Stack | Operator Stack |
|:---:|:---:|
| 001 | = |

## 6. RIGHT-HAND PROCESSING

We now begin processing what is commonly called the *right-hand side* of an assignment statement. The coding of the compiler knows that the symbol after = must be a field name. It recognizes that the single symbol B is a field name because B is surrounded by = and +. It searches the symbol table for B, does not find it, and places B on the table. Processing for + and C is performed in a similar manner. At the end of the first statement the symbol table, operand stack, and operator stack will look as pictured below. The ; delimiter signals the end of the statement.

Symbol Table

| Address | Field Name | Size | Mode | Relative Address |
|---------|-----------|------|------|------------------|
| 001 | A | 1 | Binary | Unassigned |
| 002 | B | 1 | Binary | Unassigned |
| 003 | C | 1 | Binary | Unassigned |

Operand Stack (LIFO)

| |
|---|
| 003 |
| 002 |
| 001 |

Operator Stack (LIFO)

| |
|---|
| + |
| = |

## 7. INTERMEDIATE FORM

On statement end, the compiler will undertake to create an intermediate form for representing the operands and operators of the statement. We will use a form where the elements are called *quadruples*. A quadruple consists of a label used for identification (though this is not used in all quadruples), an operator designating a function, and the operands the functions require. The quadruple form will later be used for the generation of actual instructions.

The compiler builds quadruples by use of the operand stack and the operator stack. In this example, the compiler finds a + at the top of the operator stack. Somewhere in the compiler there is a table that associates the symbol + with the address of a routine that generates quadruples for an add. This routine will know that the add quadruple needs two operands. The symbol table addresses for the two operands are on top of the operand stack. The compiler will create a quadruple of the form ADD 003 002. This quadruple contains the function add and the symbol table addresses of its operands.

The compiler must now remove 003 and 002 from the operand stack and replace them with some symbol representing the result of the addition. The symbol we use is R1, standing for Result 1. The compiler now removes the + operator from the operator stack and the result is as follows:

| Operand Stack | Operator Stack | Generated Quadruple |
|:---:|:---:|:---:|
| R1 | = | R1: ADD 003 002 |
| 001 | | |

It may be that the architecture of the machine for which we are compiling has an instruction used to add binary numbers and another instruction used to add binary-coded decimal numbers. The quadruple form of the decimal add might be ADCML. The compiler knows that it wishes the ADD form because B and C are binary numbers. It knows they are binary numbers because this fact has been recorded in the symbol table.

The compiler now unstacks the = operator and executes the appropriate code for =. This coding knows that = must have two operands. The operands are R1 (the sum of B + C) and 001 (location of A on the symbol table). The compiler now adds to the quadruple list so that it looks like this:

$$R1: \quad ADD \ 003 \ 002$$
$$STORE \ R1 \ 001$$

## 8. NEXT STATEMENT

The operator stack is now empty and the compiler proceeds to the next statement. Notice that the processing of this statement (D = A*3) will differ from its predecessor in that the symbol A occurs again and will be found in the symbol table and the constant 3 must be placed in some location in memory. At the end of processing the second statement the quadruple list will look like this:

$$R1: \quad ADD \ 003 \ 002$$
$$STORE \ R1 \ 001$$
$$R2: \quad MPY \ 001 \ "3"$$
$$STORE \ R2 \ 005.$$

## 9. CODE GENERATION

Up to this point, the design of the compiler has been due largely to various design concepts in compiling and to the specifics of the syntax of the language. From this point on, however, the specific function of the compiler is intimately associated with the architecture of the machine for which the compilation is being undertaken.

The development of quadruples is a possible point at which to define the relationship between compilation and interpretation. Notice that the quadruple form is considerably more compressed than the source high-level language form because variable names have been replaced by symbol table addresses. There is the possibility of reasonably efficient interpretation and execution of the quadruple form by an interpreter. Some architectural theorists believe, in effect, that a perfect architecture for a machine would look very much like the intermediate output of a compiler.

If we are dealing with an architecture that does not comply with the requirements of the intermediate language, and if as a consequence the intermediate language is not directly executable, then the quadruple list will be used to generate specific machine instructions. Each machine instruction will contain an operation code representing a function directly performable on the machine and a set of addresses representing the variables or operand fields that are to be operated on.

The compiler faces the problem of selecting particular operation codes and the forms of addressing to be used to reference operands. For the purposes of machine code generation in this example let's assume that the machine for which we are compiling has only one operand register to which an operand of an add must be taken before adding and in which the result of an addition is stored. Such architectural features imply associated instructions to bring an operand from memory to the register, to add the value in the register to another value in memory, and to store the contents of the register in the location designated. The instructions that we will generate are LOAD, ADD, STORE, and MULTIPLY. Since there is only one register in the machine, the register reference is implied in the instructions. In architectures that have a register model with more than one operand register an important part of the compilation process is the determination of the proper use of the operand registers in order to minimize the number of LOADs and STOREs that must be executed.

We will also, for simplicity's sake, make the assumption that all addresses of operands are going to be *direct addresses*, that is, that the full and final addresses of the operands are represented in the machine instructions. Remember that we have made the assumption that each field and each instruction will occupy one memory location. The compiler will establish a *memory assignment counter* to use in assigning addresses, which will have an initial value of 00000. The addresses generated by a compiler are usually *program relative addresses*, addresses that assume that the beginning location of a program is 00000 and that all addresses are generated relative to that number. When an address is generated it is recorded for the field in the symbol table. Address assignment is trivial in this example because we are assuming that A, B, and C will occupy one computer location. It is possible that more complex address assignments would be required for certain machines and certain representations of A, B, and C.

Remember that we are using the convention that a machine instruction is represented by a form at the assembler language level. At the assembler level an instruction to add two values may well be written as ADD. As we saw in Chapter 1, the operation code for ADD in the true machine architecture will have a unique binary value that may have no similarity to the ADD value. In addition, the actual order of fields of an instruction may be different at the true machine and at the assembly language level.

The compiler will first process the quadruple ADD 003 002. It will execute instructions in an *add generator* routine that knows that it must generate two instructions. The two instructions are LOAD, Address and ADD, Address, where Address merely represents the need for operand addresses. At this point we have not yet assigned addresses.

The generator inspects the memory assignment counter to determine the initial address of the instructions. No instructions have yet been generated so the generator uses the initial locations and adds two to the memory counter. The symbol table addresses represented in the quadruple are used for indicating B and C in the instruction. The generated instruction list now looks like this:

| Loc | Instruction | Quadruple | Statement |
|-----|-------------|-----------|-----------|
| 00000 | LOAD 003 | ADD 003 002 | A = B + C; |
| 00001 | ADD 002 | | |

The compiler now processes other quadruples until the instruction list is completed.

| Loc | Instruction | Quadruple | | Statement |
|-----|-------------|-----|-----------|-----------|
| 00000 | LOAD 003 | R1: | ADD 003 002 | |
| 00001 | ADD 002 | | | |
| 00002 | STORE 001 | | STORE R1 001 | A = B + C: |
| 00003 | LOAD 001 | | | |
| 00004 | MPY "3" | R2: | MPY 001 "3" | |
| 00005 | STORE 005 | | STORE R2 005 | D = A*3; |

The labels R1 and R2 in the above quadruple list do not refer to or imply registers. The R should be read as "result." The Rn form in the quadruple provides a way of indicating that one of the operands of a quadruple is the result of a previous quadruple.

In the above generated code the reader will notice that A is stored and immediately retrieved. Since this is a register machine, this storage and retrieval might be suppressed by a part of a compiler called an *optimizer* that might reform the code in order to reduce it to the following:

| Loc | Instruction |
|-----|-------------|
| 00000 | LOAD 003 |
| 00001 | ADD 002 |
| 00002 | MPY "3" |
| 00003 | STORE 005 |

This could only be done, of course, if the compiler recognized that there was no other reference to A than the one on the right-hand side of D = A*3. For this purpose, some symbol tables contain references to all the statements referring to a particular variable. If another reference to A is found, then the optimizer would leave the STORE, but would still be able to suppress the load.

At the end of generating instructions for the arithmetic statements, the compiler must react to the statement END. We assume that an END statement in a BEGIN/END block causes an unconditional branch whose address is provided by a loader at load time. This unconditional branch is added to the code shown below. The convention that provides for this branch allows the compiler to generate relative addresses for scalars and constants contiguous with the generated code.

It is now necessary to assign addresses to A, B, C, D, and the constant 3. In the unoptimized list of instructions, the last instruction was allocated relative memory location 00005. The generated BRANCH for END was placed in 00006. The first available memory location is therefore 00007. The value of the counter is stepped as each instruction is inspected to determine its need for an operand address. Stepping through the instruction list, the compiler generates:

| Loc | Instruction | Statement |
|-----|-------------|-----------|
| 00000 | LOAD 00007 | |
| 00001 | ADD 00008 | |
| 00002 | STORE 00009 | A = B+C; |
| 00003 | LOAD 00009 | |
| 00004 | MPY "3" | |
| 00005 | STORE 00010 | D = A*3; |
| 00006 | BRANCH XXXX | END; |
| 00007 | 0000000 (Values of B) | |
| 00008 | 0000000 (Values of C) | |
| 00009 | 0000000 (Values of A) | |
| 00010 | 0000000 (Values of D) | |

Each time an address is developed it is recorded in the symbol table. This recording allows any debugging effort to work backward through the symbol table to find the high-level name of a field if the symbol table is available at execution time.

One detail not attended to is the imbedded constant 3. If the architecture has a form of MULTIPLY instruction that can use its address field to contain the value 3, that form can generally imbed a constant into the instruction rather than address the constant in another location. Then when a constant is used such an *immediate* MULTIPLY instruction might be generated. If, however, that form does not exist, or the constant cannot be imbedded, an address must be generated for the constant, the constant placed in that address by the compiler as if it were an instruction, and the address placed in the MULTIPLY instruction.

This completes generation for this program of two assignment statements. We have not addressed how B and C receive values but assume that values are placed in B and C by some external piece of program. Also please remember that the actual instructions will have binary values for operation codes that will not read LOAD and so on.

In a more realistic compilation effort on a more complex architecture, the compiler would have to concern itself with the optimization of operand register use, with more complex operation code selection criteria, and particularly with more complex choices about addressing formats and addressing conventions.

## 10.  POST-COMPILATION PROCESSES

For simplicity's sake we assume that the compiler can hold the complete program under compilation in its own space while it is compiling. As the compiler's last act it does an output operation that writes the program to a disk. The operating system will record in a catalogue the location on disk of the compiled program.

The compiler notifies the operating system that it is completed. The operating system then regains control of the machine and notifies the programmer of completion. The programmer may respond by keying in RUN.

The operating system now starts a *loader* that will bring the compiled program into memory for execution. The loader knows what memory locations are being used by programs that must always remain in the machine. Such programs are typically part of the operating system. If permanent programs occupy locations 00000 to 00999 in the machine, the loader will *relocate* the compiled program by adding a constant of 01000 to each address. In some architectures, the loader can load an address register with the value 01000, and that value will be added by hardware to all program addresses. The loader then places the instructions and the constant in the adjusted addresses. At the end of the load it transfers control to location 01000, the location holding the first instruction of our compiled program. The reader is reminded of the representation in Figure 2 of the process of compilation and loading.

## 11.  FINAL COMMENT

This description of the compilation process is naive in many respects. We have offered it here in the belief that this process of transforming a high-level language program into machine code establishes a conceptual link between what the reader knows, his programming language, and what the computer understands and responds to.

## QUESTIONS

1. What is a symbol table?

2. How does a compiler recognize an assignment statement?

3. What is the use of stack concept in compiling?

4. What is a Quadruple?

5. How would the code generated by A = B + C; D = A*3; look if the machine had an ADD instruction with three addresses, one for each operand and one for the result.

6. In your opinion, are the three address machine codes in some sense "better"? Why?

# Chapter 3

# Data Coding
# and Reference

## 1. DATA DECLARATION

Part of the process of programming is describing the attributes of data that will be manipulated by a program. Various programming languages differ in the exact manner in which this is done, but all provide a facility for the definition of data units and almost all compilers contain default assumptions about the characteristics of data if no declaration is made.

Thus, using PASCAL, one may declare the fields and structures to be referenced in the procedural sections of the program:

> VAR a: integer
> VAR b,c,d: real
> VAR e, f, g: char
> VAR y: ARRAY [Boolean] OF char

The declaration provides us with information about how data are to be coded, how they must be allocated memory space, and, in many architectures, what instructions are needed to manipulate the data.

Depending on the size of the addressable unit of a memory and on the representation of data, some data units will be assigned to one memory location, others may be spread across several memory locations, and yet others may be packed into the same memory location.

## 2. DATA MODES AND REPRESENTATION

Data in a computer are represented by various applications of the base 2 numbering system. This system provides only two unique digits, 0 and 1, from which all values of numbers may be built exactly as they are in base 10. The binary equivalent of the base 10 number 456 is 111001000. The value 456 is understood by interpreting the string of 0's and 1's as follows:

### Binary Representation

| $2^8$ | $2^7$ | $2^6$ | $2^5$ | $2^4$ | $2^3$ | $2^2$ | $2^1$ | $2^0$ |
|-------|-------|-------|-------|-------|-------|-------|-------|-------|
| (256) | (128) | (64) | (32) | (16) | (8) | (4) | (2) | (1) |
| 1 | 1 | 1 | 0 | 0 | 1 | 0 | 0 | 0 |

The value of a number is derived by observing the presence or absence of 1 in a particular position, called a bit. The sum of all positions with a 1 is the decimal number 456. The process of counting values in power positions is, of course, exactly the way we understand 456 to be 456 in the decimal system.

### Decimal Representation

| $10^3$ | $10^2$ | $10^1$ | $10^0$ |
|--------|--------|--------|--------|
| (1000) | (100) | (10) | (1) |
| | 4 | 5 | 6 |

This tells us that there are four units of the 100 value, five units of the 10 value, and six units of the digit value in the number.

The representation of the number 456 in terms of 1 bits in a single stream of bits, as shown above, is called *pure binary* representation. In an architecture with a 12-bit *word* (each address in memory represents a collection of 12 bits), the pure binary representation of 456 would be 000111001000.

There are other ways to represent values in a computer system, though they are all variations on the uses of the base 2 numbering system. One form of value representation is called *binary-coded decimal*. In binary-coded decimal representation each decimal digit is represented by a set of bits that gives only the value of a specific digit. Thus 456 would not be representable in a string of 9 bits, as above, but by three sets of bits as follows:

| Decimal Digit | Binary-Coded Representation |
|:-------------:|:---------------------------:|
| 4 | 0100 |
| 5 | 0101 |
| 6 | 0110 |

A 12-bit word representing 456 in binary-coded decimal would be 010001010110. Each set of 4 bits is to be interpreted as a separate digit. Each bit position in each 4-bit set has binary values of $8(2^3)$, $4(2^2)$, $2(2^1)$, and $1(2^0)$, respectively. The first 4-bit set represents the value 4, the second the value 5, and the third the value 6. The actual representation used in this example is a form called packed hexadecimal.

The reader will notice that the pure binary value of the coded string for 456 is not 456 at all. Thus when the 12-bit string representing 456 in coded decimal is brought to the arithmetic or output units of a processor, there must never be any ambiguity about the coding and representation scheme.

The example of binary-coded decimal given above is only one form of a variety of coding forms used in various architectures. Architectures may represent all values in some form of coded decimal. In such machines the representations used by the I/O devices (printers, punches, terminal keyboards, video tubes, etc.) are used for arithmetic and comparative operations as well. There are also architectures that do computational and logical operations only in pure binary. Such machines must convert pure binary representation to coded decimal when it is necessary to output data. Conversion programs, sometimes aided by special conversion instructions, are common to computing.

## 2.1.  Pure Binary Characteristics

Pure binary is frequently used for the representation of control information and instructions. In an instruction the bit positions used for addresses are commonly interpreted as pure binary numbers. Pure binary is also frequently used for data fields where the range of values is more effectively represented than in binary-coded decimal and where accuracy requirements may exclude the scientific notations discussed below.

The range of values that can be conveniently represented in pure binary depends on the size of data structure that can be handled by the arithmetic/logic units of a processor. This is frequently represented in the architecture by the number of bits that can be represented in an operand register. Architectures that have 16-bit registers can represent values roughly $(2^{16}) - 1$ in that space, those with 32-bit registers can represent values up to $(2^{32}) - 1$.

Various methods are used to represent positive and negative values and these affect the range of values that can be represented. Some machines take the high-order bit and treat it as a sign bit so that the absolute value of a word is represented by all but the sign bit which shows whether the value is plus or minus. A common coding technique for binary values is the use of either 1's or 2's complement representation. In these schemes a negative is represented by its complement or its complement plus one. Programs that are being recompiled from one machine to another may find arithmetic errors occurring because of differences in the schemes used to represent negative values and consequent subtle differences in the absolute value range of the machine.

Value Representation in a 16-Bit Register

| Bit Stream | Value | Form |
|---|---|---|
| 0000000000000111 | 7 | POSITIVE VALUE |
| 1000000000000111 | −7 | SIGNED VALUE |
| 1111111111111000 | −7 | 1's COMPLEMENT |
| 1111111111111001 | −7 | 2's COMPLEMENT |

Addition of 8 to −7 in 2's Complement

| Bit Stream | Value | Form |
|------------|-------|------|
| 0000000000001000 | 8 | POSITIVE VALUE |
| 1111111111111001 | −7 | 2's COMPLEMENT |
| 0000000000000001 | 1 | POSITIVE VALUE |

Most machines have a capability for treating values as *logical* values, unsigned streams of bits representing the high-level language concept of a boolean variable. Logical values are often used to represent series of on or off conditions where each bit of a word represents a particular switch or condition of data.

## 2.2.  Binary-Coding Schemes

The binary-coding form discussed above is a form called *packed hexadecimal*. There are various schemes involving 4, 6, 7, or 8 bits and there are various codings for the different collections of bits. Forms other than packed decimal, which has the intent of representing only numbers, typically divide the bit collection into two major sections, a numeric value and a *zone* section which in the 8-bit form each have four bits. The zone bits are used to indicate what interpretation should be made of the numeric bits. Thus the character B might be represented as 1100 0010. The number 2 might be represented as 1111 0010. The distinction between them is made because of the zone bits. The specific coding scheme used determines the specific character set, the collating sequence, and the decimal number representation of a processor. The most widely used coding schemes are ASCII and EBCDIC (in the example above), names of two particular binary-coded decimal standards.

Clearly 4 bits is more than one needs for representing decimal digits 0 through 9. Unfortunately there is no set of binary bits that represents this range of values. A set of three binary bits can only represent up to the value 7, a set of four can represent up to the value 15. The interpretation of bits in sets of three is called *octal* because it represents digits of the base 8 number system, 0 through 7. The interpretation of bits in sets of 4 is called *hexadecimal*.

## 2.3.  Floating Point

Another form of value representation is *floating point*. Floating point numbers are computer representations of scientific notation where a number is expressed as a value times an integer power of an understood base. The number 12340, for example, can be expressed as $10 \times 1234.0$, $100 \times 123.40$, $1000 \times 12.340$. Thus:

| Power | Multiplier Value | Mantissa | Value |
|---|---|---|---|
| $0(10^0)$ | 1 | 12340 | 12340 |
| $1(10^1)$ | 10 | 123.4 | 12340 |
| $2(10^2)$ | 100 | 12.34 | 12340 |
| $3(10^3)$ | 1000 | 1.234 | 12340 |
| $4(10^4)$ | 10000 | .1234 | 12340 |
| $5(10^5)$ | 100000 | .01234 | 12340 |

The use of the form is to provide a notational convenience for the representation of very large or very small numbers. The representation of scientific notation on a computer is floating point. The form is to combine into a single word an exponent part representing a power and a value part representing a base value. The computational value is adjusted by the power represented in the exponent part. The form is useful to represent numbers that might not be representable directly in the value range of the computer. Precision is commonly lost.

Systems differ in the conventions used to represent floating point values. The value portion may be in pure binary or in some other form of decimal coding, the power portion may be adjusted by some constant to make computation more efficient and avoid sign changes for values less than 1. Some systems offer an option of *double-precision floating point* which extends the size of the power portion or the value portion or both. Thus in the UNIVAC 1100's a single 36-bit floating point word (the unit represented by a single 1100 address and brought to a 36-bit operand register) has a power portion of 8 bits and a value portion of 28 bits. A double-precision floating point word, however, has an 11-bit power portion and 61-bit value portion spread across two UNIVAC 1100 words.

## 3.  ADDRESSABLE UNITS

The bit strings that represent either pure binary or binary-coded values are organized in a processor and memory into addressable units of information commonly called words. The memory, we have seen, is an addressable vector of locations, each with a unique address.

The addressability of a programming language is limited to abstract concepts of a data item that may be a variable, a collection of elements such as an array, or a more complex collection in a more general structure. The addressability of a machine program is to memory locations.

Significant characteristics of an architecture are the size of the basic addressable unit, the number of bits that are contained within a referenced memory location, and the variety of addressing forms that may be used. Although there may be instructions that can cause the movement of multiple memory locations to and from the processor and its memory, and, for the selection of particular bits, even from a memory location, the idea of a basic unit of addressing is fundamental to all architectures.

### 3.1.  Basic Word Size

The size of a word varies from architecture to architecture. The current range is from 8 to 64 bits in commercially available general-purpose processors. A single address, depending on the definition of word size in the architecture, can then contain as few as 8 or as many as 64 bits.

Many processors have a range of addressable units ranging from 8 to 64 bits and the concept of a single word size becomes somewhat vague. The basic addressable unit is 8 bits (the universal byte), but an address in a particular instruction may refer to collections of bits that involve 1, 2, 4, or 8 basic 8-bit (byte) locations. The number of bytes to be moved is coded into the operation code of the instruction.

The difference between moving a 64-bit structure within the concept of a single 64-bit word or moving a 64-bit structure as eight 8-bit words is reflected in addressing structure and address arithmetic.

| 64-Bit Word | | 8-Bit Word (Byte Addressable) | |
|---|---|---|---|
| Address | Value | Address | Value |
| 00001 | A | 00000–00007 | A |
| 00002 | B | 00008–00015 | B |

Thus in generating code for the 64-bit word machine:

```
B + A
LOAD A      LOAD 00001
ADD B       ADD 00002
```

For the 8-bit word machine:

```
LOAD Longword A      LOADL 00000
ADD Longword B       ADDL 00008
```

Here "Longword" is part of the operation code and indicates that 8 bytes are to be moved. Actually, most architectures that use the multiple byte form address from the top rather than the bottom location, so that the instructions would look like:

```
LOADL A      LOAD 00007
ADDL B       ADD 00015
```

In addition, if the LOAD and ADD instructions are 32 bits long, they might both be represented in a single 64-bit word. On a byte addressable machine, if the LOAD instruction starts in location 1000, the ADD instruction starts in location 1004.

Despite the basic byte addressing nature of many architectures, the terminology "word" as an addressable unit is commonly used. The word size is commonly perceived to relate to the size of an operand register, so that if operand registers are 32 bits the word size is considered to be 32 bits even though smaller and larger structures can be addressed to partially fill registers or to affect more than one register per operation.

If a word size is defined to be 32 bits, or 4 bytes, then terminology such as *halfword* and *doubleword* is used to describe 16-bit or 64-bit structures defined as 2 or 8 bytes. Some care must be taken here because some architectures have a defined 16-bit word so that a doubleword is 32 not 64 bits.

When halfwords, words, doublewords, and so on are addressed as multiple bytes the concept of a boundary often exists. A boundary is a rule that the larger data structures can only begin at certain places, for example, in even numbered locations. If a halfword is defined on an odd numbered location, then, depending on the organization and implementation, the address will be considered illegal or a performance penalty may be paid.

## 3.2. Address Definition

Some architectures have different addressing forms for different data representations. Thus word addressing would be used for binary or floating point values, byte addressing for character values. In these architectures, instructions that manipulate binary values may have different addressing conventions than instructions that manipulate character values.

For example, a field defined to be a character string is frequently represented in a collection of bytes, each of which contains one character of the string. Reference to the full string involves an instruction that provides the starting location of the string and the number of consecutive bytes that should be transferred. Alternatively, a starting address and ending address may be provided. However, a word-oriented instruction addressing multiple bytes would provide only the starting address; the number of bytes would be implied by the operation code.

Frequently the meaning of an address is related to the operation associated with the address. Thus an address 0020 associated with an instruction like Insert Character means use the byte at 0020, but 0020 associated with ADD HALFWORD (add a 16-bit value) means use the 16 bits in the bytes at 0020 and 0019 as a value whose most significant digit is the first bit in byte 19 and whose least significant digit is the last bit in byte 20. The processor infers the necessity of moving 2 bytes from the instruction rather than from the existence of a starting address and an ending address or a starting address and length.

Architectures frequently have an ability to manipulate strings of bits that are smaller than words or that span across words. Data addresses are organized to provide the definition of a partial word to be brought to a register, or bits in the instruction show that only partial word data are desired.

## 3.3.   Word Sizes and Processor Type

The word size of a machine is somewhat related to its price/performance class. There is a tendency for high-performance scientific machines to have architectures with larger word sizes. The reason for larger word sizes in scientific machines is to provide greater precision for binary computation and greater number ranges for floating point.

One of the dramatic tendencies in computing is for smaller machines to have increased word sizes. Where 16-bit words were standard until recently, a larger and larger population of small machines now have 32-bit word sizes.

The architecture word size does not necessarily mean that the word size is the unit of transfer between processor and memory. A processor with a 32-bit architecture does not necessarily actually manipulate 32 bits at a time either in the processor or at the processor to memory interface. The actual movement of 32 bits may be accomplished by the sequential movement of 8 or 16-bit structures supported by the organization. A frequently found sequence is for a 16-bit architecture to be extended to 32 bits and then, in a later model, for the underlying organization of the machine to be extended to really manipulate 32-bit structures as units on single references to memory. This is an example of the divergence of architecture and organization.

## 4.   NONSCALAR REPRESENTATION

A simple elemental field like a character string or a scalar (single element) variable will have direct physical meaning to a machine. It will be addressed by a single instruction and manipulated as a unit.

Not all the conceptual structures that can be declared in a programming language have direct physical meaning on a machine. Thus the declaration of an array is a declaration that is not directly representable. Similarly, complex record structures are not directly representable. These multielement variables are represented in the machine as sets of individual variables to which access is algorithmically defined by sequences of instructions operating on consecutive locations or on sets of pointers that relate elements of a data structure to each other.

The statement A(I) = B(I)*C(I) in a DO (or FOR) loop references the Ith element of an array structure allocated in multiple locations in memory. The compiler must provide code that will dynamically develop addresses for each element when referenced. The ability of some languages to represent array manipulation without indexing, for example, by declaring A, B, and C to be arrays and then stating A = B × C, does not change the need for the compiler to allocate the arrays across multiple memory locations and probably generate code to address individual elements.

The hedge implied by the "probably" is due to the emergence of architectures that have the ability to manipulate vector or array structures more directly. The result is that an operation on these elements can be truly represented by one instruction backed by a complex organization that will move elements as a group, do element operations almost in parallel, and store results in memory locations as a group.

In addition, some architectures have means of representing the existence of arrays and more complex data structures directly in the architecture by *descriptors*. A descriptor is a specialized word in the architecture that provides a description, for example, of the size of an array, its coding, and its starting and ending positions. The descriptor word controls the manipulation of an array. The descriptor element is assigned a location in memory and the compiler provides the address of the descriptor element to instructions referencing the structure. Addressing individual elements and testing for boundary elements is done by the architecture using the descriptor element whenever the data structure or an element within it is referenced.

| Location Instruction | Location Descriptor |
|---|---|
| 0000 LOAD R1 0300 | 0300 AD, BIN, 500, 5000, X2 |

The above shows a reference to an array. The compiler has generated into the LOAD instruction the address of an array descriptor word located in location 0300. This descriptor word indicates, in the informal coding used here, that the word is an array descriptor word (AD), that the array is a binary array (BIN), has 500 elements, and begins at location 5000, and that the contents of Index Register 2 are always to be used to address an individual element.

Many architectures have words in special form with special uses. The presence of the word in a particular location or register or, as above, with a particular identification of its function coded into it identifies the word and its use. Thus there may be such things as *indirect address control words,* and *addressing limit control words*, with unique and special formats. We will discuss this kind of thing as we progress.


## 5. SELF-DESCRIBING DATA

In the last section we discussed the use of a descriptor word to represent a complex data structure in an architecture and the existence of specialized control words. We have also mentioned that the data representation of a variable and the interpretation of its address is often inferred from an operation code that is manipulating the data. In some real and proposed architectures the idea of describing the attributes of data directly, by using the data, rather than inferring them from operations on the data represents a generalization of the concept of descriptors. The architectural notion is sometimes referred to as *self-describing* data or *tagged data*. A feature of some architectures is the association of a set of control or *tag* bits with addressable units of memory. These bits show that the addressable unit is a particular type of data structure and control the use of the addressable unit.

Thus tag bits may show that the addressable unit is a floating point word and may be involved only in floating point operations. Similarly a word may be tagged as having no assigned value so that an attempt to use the addressable unit in a calculation will produce a processor error. Other designations identify instructions (which cannot be changed at all), specialized control words, pointers (containing addresses to other data units), participation in an array, and so on. These tag bits cannot be directly manipulated by a programmer writing applications, but they may be manipulated by special portions of the operating system. Sometimes the specialized structures have sizes that can be used only for that function. Examples of the use of coded specialized words exist in some Burroughs architectures, where a set of bits in each word indicates the special use of that word as an instruction word, an address, and so on.

Such tagged architectures have as their goal the elimination of difficult-to-find execution bugs that result from the inadvertent addressing of undefined variables, instructions, or words of inappropriate data representation.

## QUESTIONS

1. What is coded binary decimal?

2. What is floating point form?

3. What is a byte?

4. What is a word?

5. What is the use of the concept of tagged architecture?

Chapter 4

# Register Organization

## 1. REGISTERS

When introducing the elements of an architecture in Chapter 1, we mentioned that the register model of an architecture was a central feature and introduced some basic ideas about register use. Chapter 1 introduced the notion of operand, index, and control registers. Chapter 2 discussed compiling a program and generating machine code for an architecture using the *accumulator model*, that is, architecture having only one operand register to which one operand must be brought before an arithmetic operation can be performed. At that time we suggested that there were architectures that had more than one arithmetic register.

In this chapter we discuss the use and organization of registers more completely. We pay particular attention to operand registers and leave fuller discussion of control registers and addressing registers to the chapters in which they are used.

Registers have many uses in an architecture. For the computation and manipulation of data, they provide a means of referencing data elements in a kind of addressing shorthand. Thus if an operand in a register is in an architecture with a *register file model*, a reference to the operand may be made by designating register 1, and so on, rather than using a full memory address. Since the population of registers will be considerably smaller than the population of memory locations, fewer bits will be required to represent a register designation.

Registers may be used not only to hold operands but to participate in the formation of addresses in memory. Indexing is an example of this. Besides indexing, there are other addressing formation functions involving registers and values in an instruction. A register may be designated as a *base register* holding an address in memory to which an instruction address is added as a displacement to form an address. A register may be designated as holding a full address so that a reference in memory may be made by an instruction that has only space enough for a register designation.

Another major function of registers is to hold information about the control of the machine. Values that indicate the next instruction to be executed, conditions found in the data, the addresses that a program may refer to, and so on, are commonly placed in various control registers.

## 2. REGISTER AND REGISTER MODEL CHARACTERISTICS

Registers have a characteristic of addressability; they may be referred to from an instruction. Operand registers are commonly (but not universally as we shall see) referenced in an instruction. Some control registers may be explicitly referenced from instructions. Architectures differ in the way that they permit instructions to refer to control registers. There are registers that cannot be explicitly addressed from an instruction but whose use is implicit in the control of the processor. There are also registers that can be referenced only when the processor is in a control state (see Chapters 1 and 9), indicating that instructions that change the basic status of the machine in such a way as to change such things as addressing ranges may be performed.

An important characteristic of registers is their size in bits. In an architecture, register size is usually synonymous with word size. The size of registers may, however, vary within an architecture, with certain control registers differing in size from operand registers.

A characteristic of a register model is the number of registers that exist and the number of different types of register. There are *general-purpose register* architectures in which a single set of registers may be used for operand registers, index registers, and so on. The particular use of a register is determined by where in the instruction word the register is referenced.

| Op Code | Op Reg | Index | Memory |
|---------|--------|-------|--------|
| ADD     | 2      | 2     | MEM    |

If performed on a general-purpose register set, the above instruction will not only add to the contents of register 2 the contents of MEM + (INDEX 2), it will use the initial contents of register 2 to form the address of the operand in memory. This may be pretty tricky coding, but it implies a benefit for the general register set organization: There is no need to compute addresses and move them to an index register.

An architecture may have disjoint register sets where sets of registers have special functions. In an architecture with disjoint index and operand registers, the above instruction would in fact be referring to two different registers, an operand register 2 and an index register 2.

In the 370 generic architecture, for example, there is a general register population which consists of 16 general-purpose registers, each of 32 bits, that may be used as operand registers, index registers, and registers to hold addresses or parts of addresses. The specific employment of a particular register is indicated by where the register is referenced in the instruction format. Instruction formats are defined so that specific bit positions are set aside for operand, index, and address register specification. That architecture also includes a set of floating point registers that are 64 bits in length and may be referenced only by the floating point instructions. That is, the op code of an instruction determines which register set to use.

## 3. REGISTER OPERAND REFERENCING

In an architecture using the pure accumulator model, only one operand may be addressed in a register. In such architectures there is commonly an additional register used to handle products and remainders that exceed the size of the accumulator. (The multiplication of two values of bit length n will result in a product greater than n requiring additional space. Similarly, division commonly results in a quotient and a remainder larger in bit size than the divisor or dividend). The designation of an accumulator or extender register is made implicit in the operation code. Thus an instruction ADD MEM1 appears to have only one operand because it explicitly references only the operand in MEM1. The other operand is implied by the operation code and is in the accumulator. ADD will add the value in the accumulator to the value in MEM1. There may be direct manipulation instructions on the extension register, commonly called an MQ register (e.g., an instruction LOAD QUOTIENT).

In register file architectures it is possible to address multiple operands in registers. Thus instructions may exist that have either two or three operand addresses, all of which are register addresses.

ADD Register 1, Register 2, (Register 3)

The above instruction designates that operands are in registers. The parentheses indicate that some architectures provide for the designation of a register in which to place the result, while others place the result in one of the operand registers.

Register file architectures must also have instructions that load and store values into and from registers. In general they have a class of instructions that specify a register address and a memory address to provide for moving values between registers and memory.

The major benefit of the register file model is the provision of space where values can be developed and manipulated without reference to memory. In complex computations it is frequently convenient to develop values in separate registers then combine the values.

Register file architectures frequently have a concept of multiple register manipulation with a class of instructions that load and store from consecutive registers. An instruction like LOAD MULTIPLE or STORE MULTIPLE may specify one register address but manipulate the contents of the referenced register and one or more other registers. Some architectures allow more than one register designator in the multiple type instructions in order to define a subset of registers to be involved in an operation. There are sometimes constraints about what register sets may be defined in multiple register instructions. For example, the first register must be even numbered.

Operand register sets may have some operand register use specialization. For example, in both the IBM 370 generic architecture and in the CRAY-1 there are operand registers of special function. In the IBM architecture the register distinction is between registers used for operands associated with binary arithmetic and registers associated with floating point arithmetic. The register address must be interpreted within the context of whether the instruction is a binary or a floating point instruction.

The CRAY-1 uses different sets of registers to hold operands for scalar values (values representing variables of one element) and vector values (values representing elements of vectors or arrays). The instruction code indicates which set of operand registers is addressed from the operand register positions of the instruction.

Many architectures with the register file model also permit an extended accumulator model format. In an architecture with many operand registers, the accumulator model involves the use of arithmetic instructions that designate an operand register and a memory address. In pure accumulator model form an ADD sequence looks like:

LOAD      MEM1

ADD      MEM2

STORE      MEM3

In pure register model form an ADD sequence looks like:

LOAD      R1, MEM1

LOAD      R2, MEM2

ADD      R1, R2

Accumulator model form in an architecture with many operand registers would look like:

LOAD      R1, MEM1

ADD      R1, MEM2

STORE      R1, MEM3

This accumulator form is like the pure accumulator form in that an ADD instruction can refer to memory. It is unlike it in that the specific register must be stated rather than implied by the operation code.

A compiler for such an architecture must determine which of the forms to use if both are permitted. In IBM/370 generic architecture both forms are permitted. In CDC Cyber architecture only the register file form, requiring two register loads, is permitted.

Some architectures, called *storage-to-storage* architectures, do not have registers at all. The nature of these architectures will be discussed in the chapter on addressing conventions. There is one additional register model, the *stack model*, which will be discussed later in this chapter.


## 4. OPERATIONS ON REGISTERS

The basic operations on a register include:

1. Loading the register with a value.
2. Storing a value from a register.
3. Using a value in a register as an operand in a computational or logical instruction.
4. Comparing a value in a register to another value.
5. Moving the contents of a register (shifting) in order to scale values or for some other reason.
6. Changing the sign of the register without changing the value in the remainder of the register.

The specific use of a register is determined by the operation code of an instruction.

A load of a register involves the transfer of a full word from memory or from another register. If the register is 36 bits, then a 36-bit value is placed in the register and the previous contents of the register are completely destroyed. There are some variations even in placing full values in registers. It is possible to designate that, regardless of the sign value of the word being loaded, the sign of the register should remain positive.

In many architectures there are diverse ways to partially manipulate the contents of a register, inserting streams of bits in designated portions and leaving the remainder of the register unchanged. The preparation of values by the insertion of $ signs, decimal points, and so on may include this feature.

There are also various ways of bringing values that are smaller than the size of a register to a register. Thus a 16-bit value may be loaded into a register that is 32 bits with an instruction that designates a halfword operation. The value is located in the low-order bits of the register, and the remainder of the register is set to 0 or sign extended. In architectures that use complements for negative values (see Chapter 3), a negative value will have a 1 bit in its high-order position. When loading a 16-bit negative value into a 32-bit register, it may be desirable to extend the 1 bit across the upper unused 16 bits of the register so that arithmetic will cause carries at the proper times.

One of the word-oriented architectures that has a very general scheme for placing and storing values smaller than register size is the UNIVAC 1100. Portions of a 36-bit word of length 6, 9, 12, or 18 bits can be selected for loading into a 36-bit register. The values are right justified in the low-order bits of the register, and sign extension may or may not be specified. Other architectures, among them 370 generic and INTEL 8086, provide registers of fixed sizes but indicate in the coding of an instruction that register contents are only to be partially affected by a register/memory move. The 370 generic contains the concept of loading and storing 8, 16, or 32-bit structures. The 16-bit registers of the Intel 8086 can be treated as two 8-bit half-registers, each of which can be uniquely addressed if desired.

The use of registers in forming memory addresses will be discussed in Chapter 5.

### 5. MULTIPLE REGISTER SETS

Processors that are running multiprogramming operating systems experience the phenomenon of *task switch* at frequent intervals. A task switch is the replacement of the program running on a processor with another in such a way that the suspended program can be resumed at the point of its suspension. A task switch can occur as a result of the zeroing out of a clock determining the length of time a program may have continuous control of the processor; of an operating system determination that a required resource (such as an I/O record or more memory) is not available; or of a program decision that it wishes to wait for some system event to occur before continuing.

The mechanics of a task switch involve some basic hardware and operating system concepts. The actual task switch is accomplished by a program in the operating system called a *dispatcher*. This has a list of software defined tables generically described as *program control blocks*. Each of these contains information that represents the status and capabilities of a program eligible to run on the processor. The control block may contain space for register contents, address bases and ranges, and so on. The actual contents of a control block differ widely from machine to machine and from operating system to operating system.

The set of control blocks is organized into a *dispatch list* that represents the set of programs that may contend for time on the processor. On a system with a single processor, a uniprocessor, only one of the programs can actually be running at a particular time. This program has the use of any machine register and all the registers that participate in address formation are set to indicate the addressable space of this program.

In an architecture that has one set of general-purpose registers, the dispatcher does a task switch by storing the contents of all the registers into the control block in memory, loading the registers of a successor program with values found in its control block, and transferring control to a designated location in the successor program. This designated location is usually the address of the instruction following the last instruction executed at a prior running of this program. The dispatcher, when it suspends the program, stores the contents of the instruction counter into the control block.

The time required for the storage of register values from the retired program and the establishment of values for the successor program may be long if register populations are large and if the architecture does not provide much hardware support for the task switch. In some architectures all status information must be stored by the execution of the usual store register instructions and all new information established by the execution of the usual load instructions.

To speed task switch two architectural approaches have been taken. One is to improve the efficiency of register save and restore, the other is to eliminate or minimize it. Increased efficiency may be achieved by LOAD and STORE MULTIPLE instructions that manipulate sets of registers with single instructions. Some architectures go beyond this and automatically save registers by hardware whenever a task switch occurs. A form of task switch instruction attends to suspending the running program and includes the saving of registers and the setting up of a successor program including loading registers.

Another approach is to eliminate or reduce register save and restore. This can be accomplished with duplicate register sets so that there is a complete set of general-purpose or specialized noncontrol state registers for each program that may be started on the machine. In a non-IBM version of 360 architecture, for example, there are 16 sets of the 16 general-purpose registers. The number of program control blocks that the operating system will support may be limited to that number so that whenever task switch occurs it is not necessary to store and restore registers. Whenever a task switch is performed, a control register is set to point to the set of registers that holds values for the program about to be activated. An almost identical concept exists on the IBM 8100.

Some architectures do not go quite this far. They provide only a set of problem state registers and a set of registers for the operating system to use when the machine is in control state. The motivation for this is that the operating system is frequently entered for reasons that will not result in a task switch. The problem program will call for a service to be performed by the operating system or an event will occur that starts the operating system, but, when finished, the operating system will characteristically return to the program that was running. The performance of the machine can be improved by having just two sets of registers. This avoids saving and restoring whenever the processor moves from processing a problem program to running the operating system and then returns to the program that was serviced. UNIVAC 1100 architecture has this feature.

## 6. OPERAND STACKS

Registers have so far been treated as sets of independently addressable storage locations that are linked to each other only for multiple register instructions, for double precision floating points, or for quotient or product overflows. There is a class of machine that establishes a more formal relationship between registers and eliminates direct addressability to registers from a program. Register references are implied and the particular register used for an operand or address is determined by the context of the instructions. These machines are called *operand stack machines*. The first commercially available operand stack machine was the Burroughs 5000. Its various successors have retained and extended the idea of operand stacking.

Consider how the instructions below may be made meaningful.

LOAD B
LOAD C
ADD
MPY

At first we seem to have an accumulator machine because LOAD does not specify a register. However, if the code is correct we cannot have an accumulator machine because then the contents of the accumulator will be C and the loading of B will be lost. We also have a problem in that ADD and MPY seem to specify no addresses at all.

This code can be made meaningful if we transform a register file model into an *operand stack* model. In such an architecture the registers are considered to be a LIFO stack. There is a control register, called the *top of stack pointer*, that indicates what register is at the top of the stack. Being at the top of the stack means that the register is next to receive a value. At any one time, the register pointed to by the top of stack register is the one that will next receive an operand. The register just beneath the top of stack is the register that holds the last operand or result delivered from memory or from an operational unit.

Using once more the statement A=B+C*D, an operand stack machine may execute the following code. In this code we use the stack-oriented terminology of PUSH and POP in place of LOAD and STORE. PUSH means put a value on top of the stack and adjust the stack pointer; POP means take a value from the stack and adjust the stack pointer:

1. PUSH D
2. PUSH C
3. MPY
4. PUSH B
5. ADD
6. PUSH (Address)A
7. POP

This code operates in the following way:

1.  PUSH D: Place the contents of the location D in the register pointed to
    as the top of the stack. Move the top of stack pointer register up one
    register.

> Register 1   Empty
> Register 2   Empty
> Register 3   Empty > TOP OF STACK
> Register 4   D

Notice that the top of stack pointer has automatically moved from Register 4 to Register 3 after the load of D.

2.  PUSH C: The contents of the memory location holding C are brought to
    the register at the top of the stack. The population of registers now looks
    like:

> Register 1   Empty
> Register 2   Empty > TOP OF STACK
> Register 3   C
> Register 4   D

3.  MPY: A multiply operation requires two operands. The two operands
    required are in the registers just below the top of stack register. Without
    needing explicit addressing, the instruction will find its operands avail-
    able to it in the registers and can use them to do the multiply. The product
    will be returned to the stack, and the top of stack pointer will be appro-
    priately moved. The stack now looks like:

> Register 1   Empty
> Register 2   Empty
> Register 3   Empty > TOP OF STACK
> Register 4   C*D

4.  PUSH B: The contents of the location holding B are brought to the
    register at the top of the stack. The stack now looks like:

> Register 1   Empty
> Register 2   Empty > TOP OF STACK
> Register 3   B
> Register 4   C*D

5.  ADD: This operation also requires two operands. The addends are in
    the registers immediately below the top of stack pointer. The add takes
    them, performs, and returns a result to the stack.

Register 1   Empty
Register 2   Empty
Register 3   Empty > TOP OF STACK
Register 4   C*D+B

6.  PUSH (Address)A: The location A does not contain an operand but is
    the address of the location where the computed value will be stored. The
    compiler must generate this address, mark the generated value as an
    address in some way, and make it available to the PUSH instruction as
    either an immediate value or a constant stored in a location used by the
    PUSH. The stack now looks like:

Register 1   Empty
Register 2   Empty > TOP OF STACK
Register 3   Address A
Register 4   C*D+B

7.  POP: This is also a two operand instruction, requiring a value to store
    and a location to store into. These values are available on the stack and
    the storage of the computed value into A can be accomplished without
    addresses in the instruction. The stack is now empty and the top of stack
    pointer is set at Register 4.

The top of stack pointer register contains the address of the register at the top
of the stack. In this example it is a transparent register since it is automatically
manipulated by hardware and not referenced by an instruction. All operand
stack machines contain instructions that can change the value of the top of
stack pointer directly. These instructions allow programs to have some direct
control over the determination of the arithmetic register at the top of the stack.
  Readers are probably suspicious enough to be alarmed by the notion of
"can," since it always implies situations in which one MUST. And it is this area
in which objections are raised to operand stack machines. Some people think
there are too many instances in which the stack must be directly manipulated
for stacks to be efficient. These instances include the return of extended results
and other side effects of arithmetic operations.
  Notice also that the PUSH and POP operations of a stack architecture are
equivalent to the LOAD and STORE operations of a register file model. Some
architects, as we see in Chapter 12, object to the necessity for this and would like
to minimize the use of instructions that merely move data from one place to
another. They contend that loading and storing from the stack is not efficient
even though address compression may be achieved.
  Another objection to the stack comes from its basic characteristic. Given a
set of registers organized into a stack one must determine how many registers

the stack will have. If the register stack is too shallow, a mechanism for defining extensions in memory must be devised. Movements of operands that overflow the register population of the stack will result in movements that are essentially memory to memory (memory to a part of the stack defined as an overflow area in memory) and then memory to register as the stack is manipulated.

If the stack mechanism is too long, then the registers in the register file may not be effectively utilized. The number of operands in the stack will characteristically not fill the registers on the stack and the register file will scarcely be used.

We now come upon an architectural issue that is very closely related to some organizational considerations, in particular, the degree to which arithmetic operations may be performed in parallel on a high-performance architecture. This in turn relates to some issues in compiler strategy.

Consider an architecture with eight registers in a register file model. Consider the flow of statements:

$$A = B + C;$$
$$F = G + H;$$
$$R = T + W;$$

These assignment statements have the property of being unordered; no statement is dependent on a predecessor for execution. A simple-minded compiler for a stack architecture might generate:

```
PUSH B
PUSH C
ADD
PUSH (Address) A
POP
PUSH G
PUSH H
ADD
PUSH (Address) F
POP
PUSH T
PUSH W
ADD
PUSH (Address) R
POP
```

Notice that the stack is never more than two deep and that six of the eight registers are never used in this sequence. To use the register stack more intensively and for reasons of organization having to do with parallel operations between processor and memory which we explore in Part 2, a more model-sensitive compiler might generate:

| | |
|---|---|
| PUSH B | Bring B to Stack |
| PUSH C | Bring C to Stack |
| PUSH G | Bring G to Stack |
| PUSH H | Bring H to Stack |
| PUSH T | Bring T to Stack |
| PUSH W | Bring W to Stack |
| ADD | Add W + T, Remove W, T, Place W + T on Stack |
| PUSH (Address) R | Bring Address of R to Stack |
| POP | Place W + T in R |
| ADD | Add G + H, Remove G, H, Place G + H on Stack |
| PUSH (Address) F | Bring Address of F to Stack |
| POP | Place G + H in F |
| ADD | Add B + C, Remove B, C, Place B + C on Stack |
| PUSH (Address) A | Bring Address of A to Stack |
| POP | Place B + C in A |

Here we have flowed operands into the stack, perhaps taking advantage of some concurrent organizational features, using the register file more intensively. We have reversed the order of execution, but, because the statements are independent, may not care about this.

But the notion that we may have some liberty about the order of execution of independent statements raises an objection about operand stack architecture. In an architecture with eight registers and an organization and implementation that provide for potential parallel execution, the following code might be very desirable for the three-statement sequence, A = B + C, F = G + H, R = T + W:

```
LOAD R1, B
LOAD R2, C
LOAD R3, G
LOAD R4, H
LOAD R5, T
LOAD R6, W
ADD R1, R2
ADD R3, R4
ADD R5, R6
STORE R1, A
STORE R3, H
STORE R5, R
```

The flow of operands into registers may occur as with the sequence of PUSHes on the stack architecture. However, the ability of the ADD instructions to uniquely address the registers is very important. It provides an ability to execute the ADDs in parallel on a parallel organization that has some form of multiple addition functional units. This could not be easily done on a stack architecture because of the implied fixed LIFO order of the stack.

Thus we see that the fundamental nature of the stack, its LIFO quality, and the ability to suppress register addresses can become liabilities on a high-performance machine with a high degree of parallel operation. The upstate returns on this issue, as we see in Chapter 12, are by no means in. What these comments strongly suggest, however, is that there are relationships between architecture and organization where the performance goals of a design influence the choice of a register model for the architecture.

## 7. STACKS AND COMPILATION

Chapter 2 discussed the processes of compilation. We said before that the stack concept is a way of supporting the processes of compilation with some efficiency. We show the relation between operand stacks and compilation using the example statement A = B+C*D.

This statement is said to be in *infix* form. Infix means that the statement has the form OPERAND FUNCTION OPERAND. In Chapter 2 we showed a way of turning this form into quadruples during the phases of compilation. We now show a supplementary compiler process involving the conversion of the infix form to the *prefix* or *postfix* form which is more convenient for compilation.

We use the postfix form which means that the representation of operands and functions is always OPERAND OPERAND FUNCTION. The postfix form of A = B+C*D is ABCD*+=. The interpretation of this form (from the recognition of the *) is as follows:

1.  * Multiply the two operands immediately to the left of *, giving the result AB(C*D)+=.
2.  + Add the two operands immediately to the left of +. These two operands are the result value (C*D) and B. This gives A(B+C*D)=.
3.  = Store the operand immediately on the left in the operand just beyond, giving A equal to the computed value.

The generation of the transformation of infix to postfix in the compiler itself depends upon a LIFO stack and a concept of *operator strength* that tells us that in statements like A = B+C*D the C*D is always executed before the addition. In general, exponentiation is done before the division, division is done before multiplication, and multiplication before addition or subtraction whenever the notation is ambiguous.

The transformation of infix into postfix is accomplished, on an operand stack machine, in the following way:

1.  Begin.

| Input Stream | Stack | Output Stream |
| --- | --- | --- |
| A = B+C*D; | EMPTY | NULL |

2.  Find A. Place A on stack.

| Input Stream | Stack | Output Stream |
| --- | --- | --- |
| = B+C*D; | A | NULL |

3.  Find =. Pass to output stream.

| Input Stream | Stack | Output Stream |
| --- | --- | --- |
| B+C*D; | A | = |

4.  Find B. Place B on stack.

| Input Stream | Stack | Output Stream |
| --- | --- | --- |
| + C*D; | B | = |
|  | A |  |

5.  Find +. If + has greater strength than =, pass to stream.

| Input Stream | Stack | Output Stream |
| --- | --- | --- |
| C*D; | B | += |
|  | A |  |

6.  Find C. Place C on stack.

| Input Stream | Stack | Output Stream |
| --- | --- | --- |
| *D; | C | += |
|  | B |  |
|  | A |  |

7.  Find *. If * has greater strength than +, pass to stream.

| Input Stream | Stack | Output Stream |
| --- | --- | --- |
| D; | C | *+= |
|  | B |  |
|  | A |  |

8. Find D. Place D on stack.

| Input Stream | Stack | Output Stream |
| --- | --- | --- |
| ; | D | *+= |
|   | C |   |
|   | B |   |
|   | A |   |

9. Find Statement Terminator. Pop stack to stream.

| Input Stream | Stack | Output Stream |
| --- | --- | --- |
| NULL | EMPTY | ABCD*+= |

In all instances in this example succeeding operators had a higher strength than predecessors. If any had been found with a lower strength, operands would have joined the output stream at that time.

Notice that when the postfix form is generated the generation of application code as first shown in the discussion of the behavior of the stack is quite straightforward. Using a stack and generating for a stack machine, the postfix ABCD*+= leads directly to:

1. PUSH D (* and leftmost operand).
2. PUSH C (* and next leftmost operand).
3. MPY.
4. PUSH B (* with leftmost on stack and B next leftmost).
5. ADD.
6. PUSH (Address)A (= with leftmost on stack and A next leftmost).
7. POP.

The technique used to generate this code is to pass the postfix string once more. As operands are encountered they are placed on an operand stack. The operands at the top of the stack are those required for the next operator encountered. Thus, for example, C and D are at the top of the stack when * is encountered.

The instruction sequence for the stack machine may be altered to optimize performance on a stack as on any other type of machine. The organizational reasons have to do with assumptions about parallel execution, memory bandwidth, partitioning, and so on.

It is possible that during compilation or run time the accumulated operands may exceed the number of registers that form the stack. A facility for deepening the stack by logically adding locations in memory and pushing operands into these locations is common to operand stacks. A register points to memory locations that are considered to be part of the stack. The use of memory locations as part or all of a stack is common, as we shall see, when stacking concepts are used to support subroutine linkages.


## 8.  SOME EXAMPLE OPERAND REGISTER MODELS

The register model of the IBM 370 generic has been alluded to several times. In this section we show the register models of a variety of architectures of different style.

### 8.1.  CRAY-1

One of the most interesting register models is that of the Cray Research CRAY-1. The operand model is pure register file where both operands must be in a register and arithmetic instructions cannot refer to memory. The processor contains:

1.  A set of eight 64-bit operand registers for scalar variables (S registers).
2.  A set of eight 64-word (4096-bit) vector registers (V registers).
3.  A set of eight 24-bit address registers (A registers).
4.  A set of 64 64-bit scalar save registers (T registers).
5.  A set of 64 24-bit address save registers (B registers).

The operand registers (S registers) are individually addressable and are used to hold one-word scalar variables while fixed or floating type arithmetic, boolean operations, and shifts are performed.

The vector registers (V registers) contain elements of arrays and vectors when identical arithmetic, logical, or shift operations are to be performed on many elements of a vector. Each 64-word vector register group contains a group of 64-bit registers. Vector operations permit an operational unit to do functions on elements of vectors with the issue of a single instruction. The function performed is indicated in the operation code. The register address portion of an instruction refers to a particular set of vector registers, for instance, V0, V1, and so on. A supporting register called the vector length register (VL) defines exactly how many of the 64 elements of a vector are to be included in an operation. It is also possible to do functions on particular elements.

The Address registers (A registers) may be used as index registers, to hold shift count values, and control I/O. The concept of an address register is somewhat unique to the CRAY and CYBER architectures. An instruction of the type LOAD R1, Memory, actually does not refer to the S or V operand register but to an A register. There is a correspondence between A registers and S and V registers such that a reference to R1, an A register, implies a reference to a linked operand register. The memory location stated in the LOAD instruction is placed into an A register, and, as a result, a value is loaded or stored from a corresponding S or V register. This is a form of implied addressing. A registers may also receive operands from memory, send operands from memory, and send or receive data from S registers.

The B and T registers are rather unique to CRAY architecture as well. The intent of these registers is to provide a significant set of temporary storage areas where values moving to or from A or S registers may be held. Such values form a kind of *working set* of values that are not in immediate use but which are intended for use or reuse shortly. Multiple words may be moved between B and T registers and memory. In effect these registers provide a programmer with a controlled local scratchpad or operand/address buffer that permits the processor to minimize its memory references.

The register model of the CRAY-1 actually provides an alternative to *cache* designs, which we discuss later. A cache is a high-speed memory whose content is automatically controlled by hardware. The CRAY-1 takes the approach that the contents of fast memory, as represented by its enormous set of registers, is best controlled by program. It is an architectural solution, as opposed to a purely organizational solution, to the problem of high-speed processors being typically much faster than the memories they need.

## 8.2. UNIVAC 1100

The UNIVAC 1100 register model is essentially an extended accumulator model. The UNIVAC 1100 register population is called the *General Register Set* (*GRS*). The general register set is addressable as locations in lower memory or from register designation fields in UNIVAC instructions. Locations in the general register set are also used for various processor status and I/O control functions.

The fundamental registers in the GRS are the A registers (operand registers), X registers (index registers) and the R registers. Sixteen A registers are addressable from the A register bits of an instruction. Sixteen registers are addressable from the X register bits of an instruction. A registers 0 to 3, however, are the same registers as X registers 12 to 15, so that some registers may be used as either operand or index registers. Besides register overlap there is memory addressing of registers so that the contents of registers may be moved from one register to another by addressing one register using a register address and another by using a memory address below a certain value.

R registers are special-purpose registers used to hold, for example, values for a *real time clock*. A real time clock has the function of counting down periods of time and emitting a signal when the register is reduced to 0. It controls the operation of applications in real time environments when responses must be made within a certain time. Another R register is used as a *repeat count*.

The UNIVAC 1100 has an instruction that will execute instructions a multiple number of times. The number of times that an instruction will be executed is loaded into an R register. The feature is useful, when used with the index register decrement feature, to effectively write programs that will search a list of values for a particular key. If the key is not found, the search will end when the repeat count goes to 0. If the key is found, the address of the key can be derived from the contents of an index register at the time the key was located.

### 8.3.  VAX-11

The VAX-11 has a set of 16 registers, of which 12 are general-purpose and four have specific uses in support of stacking operations or subroutine linkage. One of the 16 registers is the instruction counter. This register is called the *program counter* in this architecture.

Since all operations can be performed on the VAX-11 in either registers or memory, the use of the general-purpose registers is less critical than in machines that can only do certain operations in registers. The compression of address space and the fast access to registers, however, are still important attributes of register use on this architecture.

### 9.  SOFTWARE REGISTER CONVENTIONS

We have looked at many of the ways that operand register populations can be defined. Architectures differ in the number of registers, the use of the registers, and the way the registers are related to each other.

Those architectures that have general-purpose register populations with no special functions associated with any registers may establish conventions for register use in the compilers and the programming manuals. Thus there may be a convention that certain registers are always used to pass parameters or that certain registers are always used to indicate return points. These software conventions can define the use of registers in a general-purpose register machine as effectively as if the registers were assigned special functions in hardware.

### 10.  BASIC CONTROL REGISTERS

In general we discuss control registers in connection with the specific control functions they perform. However, we introduce here some of the basic control registers to give the reader a taste of the types of function they support. The basic control registers of a machine are the *instruction counter, instruction register*, and *memory interface register*. These registers are visible to the program in some architectures that may manipulate them with instructions, but they are frequently modifiable only by automatic action on the part of control circuitry. In architectures with a *control state* (introduced in Chapter 1)—a condition of the processor that enables certain instructions that cannot be executed in normal application operation to be executed—it is common for instructions that modify control registers to be executable only in this state. This gives the operating system the ability to change machine status in ways that a usual application program cannot.

An instruction counter holds the address in the memory of the next instruction to be executed after the completion of the instruction currently being executed. Different organizations advance the instruction counter at different times in the control cycle, but the general effect is that during the execution of an instruction the instruction counter holds the address of a successor. Thus, on a machine level memory dump, taken at a point when the machine has stopped executing a particular program, the contents of the instruction counter are the last instruction executed plus the constant used to step through the memory. On a word machine the constant may be one, on a character machine the constant will be the number of characters occupied by the current instruction. Thus, if the instruction counter reads 0104 when a two-character long instruction is fetched, it will read 0106 during the execution of that instruction.

The instruction register holds the operation code and address fields of an instruction while it is being prepared for execution and actually executed.

The memory interface register is a location where the address to be fetched from memory next, or to which something is to be stored, is held for an interval of time until circuits in the memory can respond to the request for a fetch or storage cycle. In rather small machines it is also where the quantity taken from memory is placed at the end of the memory fetch cycles. Thus, on a processor fetch reference to memory, the memory interface register first holds the address and then the data.

On a store to memory, a very small processor may use the memory interface register to sequentially hold an address and then to hold the data to be placed. Register space is now becoming so inexpensive that it is common to find that even in very small systems there may be an elaboration such that there is a *memory address register* and a *memory data register*. This allows memory circuits parallel access to address and data to be stored. In certain organizations it permits some overlap between memory storage and memory fetch.

Beyond the basic registers that hold the address of an instruction, the instruction itself, and data moving to and from memory, there will be many other registers. There may be registers that control I/O flow, interrupt reporting, memory addressing management beyond what we have yet described, and so on. Additional control registers in the memory area include limit registers that define the range over which a program can address. Page and segment origin registers, used in paging machines, are described in a later chapter. Stack pointer registers, described primarily as register address pointers in this chapter, are also used to define stack and stack element areas in memory.

## 11. PLACEMENT OF CONTROL INFORMATION

An interesting aspect of architecture is the split of control function between designated control registers and locations in lower memory. It is common for a system to use the first few hundred or thousand locations of memory for special purposes. Some control information is placed in control registers and some in control locations in memory. These memory locations may hold such things as the instruction counter or the register contents of a program that has been task switched. In addition, the locations may hold various *status words* that reflect the status of the system at the end of an I/O operation, and so on. The split of function between lower memory locations and special registers sometimes seems arbitrary, especially in older architectures whose features have been extended over time.

## QUESTIONS

1. What is the purpose of including more than one set of general-purpose registers?

2. What is an operand stack?

3. What efficiency in addressing does a stack architecture provide?

4. Develop an argument indicating why stack architecture may not be necessarily more efficient for high performance.

5. What is the use of the Save class of registers on CRAY-1?

6. How may UNIVAC 1100 arithmetic registers be addressed?

**Chapter 5**

# Memory Addressing Conventions

## 1. ADDRESSING CONVENTIONS IN INSTRUCTIONS

In Chapter 3 we discussed the meaning of an address in terms of the size and representation of data elements to which an instruction might refer. In Chapter 2 we discussed compilation for an architecture that had one operand register and that completely represented addresses in the address portion of an instruction. We have been introduced to the concept of indexing, and we have seen from the last chapter that operands may be addressed in registers.

In this chapter we discuss operand register addressing more fully, complete our discussion of indexing, and explore another aspect of addressing: the various methods that an architecture may use to combine the value in the address portion of an instruction with other values to form a complete, current address.

Every instruction may be viewed as having the general form OPCODE, OPERAND1, OPERAND2, . . . , OPERANDn. Actual processor architectures will vary as to the number of addresses an instruction contains and whether addresses may be full memory, partial memory, or register addresses.

## 2. FUNDAMENTAL ISSUE

An instruction that contains multiple full memory addresses is rather long in terms of the number of bits required to hold the addresses. To reduce the size of programs in memory and the number of bits that must be moved between processor and memory to execute a program, decisions are made about how many addresses should be used in an instruction and how long these addresses should be. The constraint on reducing the size and number of addresses in an instruction is that it may be necessary to use more instructions to accomplish the same function. The increased population of instructions, even though each one may be shorter, may increase the size of programs beyond the savings in instruction size. In addition, the execution of additional instructions may take more sequential time on a machine, depending on the complexity of the organization of the machine. So there are significant and difficult decisions to be made about address forms.

An additional issue derives from concern about methods for constraining the memory locations that a program can reference to protect the security, or integrity, of information in a memory being shared (multiprogrammed) by multiple programs. Many architects feel that addressing schemes should reflect concepts of protection and limit the ability of a program to address memory to those portions where it has relevant instructions and data. Finally, some architects feel that addressing schemes should in some way reflect the structure of a program in terms of its instruction areas, data areas, relationship to subroutines, and so on. We start to address these issues in this chapter but complete our discussion in Chapter 10.

## 3. ADDRESS SIZE COMPRESSION

Let us look more closely at the problem of instruction size. Consider an architecture that has an ADD instruction that specifies the location of two operands and a place where the result is to be placed. This is called three-address storage-to-storage. If the range of addressable words in the architecture is 16 bits, then the ADD instruction must have 48 bits of address. If a 6-bit operation code is used, then the instruction will be 54 bits. This does not count any bits needed to use index registers to modify the address. In the example below we assume 2 bits associated with each address to indicate index register modification for each address, giving a 60-bit instruction.

| Op Code | Operand 1, Index | Operand 2, Index | Result, Index |
|---------|------------------|------------------|---------------|
| ADD     | 16 bits, 2 bits  | 16 bits, 2 bits  | 16 bits, 2 bits |

One way of shortening this instruction is to reduce the number of bits in each address. One way of reducing the number of bits is not to represent the full addressing range in the architecture in an instruction. Thus, although the full memory can be addressed only with 16-bit addresses, only eight address bits and two index reference bits are used in instructions for each operand. This implies that there is a register, besides the indexing registers, that holds a value that can be used to form a full 16-bit address. Coding for such an architecture, assuming an 8-bit register that holds the upper bits of an address for all instructions might look like:

LOAD BASE REGISTER "Base"
ADD OPERAND 1, INDEX; OPERAND 2, INDEX; RESULT, INDEX:

The bit length of the ADD instruction is now 6 bits for operation code and 8 bits plus 2 bits for index register designation for each address, giving a 36-bit rather than a 60-bit instruction. The LOAD BASE REGISTER instruction contains a 6-bit operation code and an 8-bit value to be put in the register. For each address, the 8 bits in the register are used to represent the eight uppermost significant bits of address and the 8 bits in the instruction to represent the least significant 8 bits of address. A 16-bit address is formed by concatenating the bits from the instruction to the bits in the register. Indexing values are then applied to this value.

The LOAD BASE REGISTER instruction may be held in a 36-bit word with 22 unused bits. However, an architecture might provide for variable instruction sizes so that this instruction might be smaller and represented in, for example, half the size of the ADD instruction, or 18 bits.

Shortening addresses will decrease the size of a program and reduce the number of instruction bits that must be moved between processor and memory, if (1) base register load instructions are shorter than the arithmetic instructions and (2) many arithmetic instructions use the same base register value eliminating frequent LOADS of this value.

The smallest form of address that may be used in an instruction is a *register address*. Thus an ADD might look like:

ADD REGISTER 1, REGISTER 2, REGISTER 3

In a register file architecture, if the processor has 16 operand registers, then 4 bits are necessary for each address, giving an 18-bit instruction. The problem is, however, that we must have instructions that place the operands in the registers and store the result from a register.

```
LOAD R1, OPERAND 1
LOAD R2, OPERAND 2
ADD R1, R2, R3
STORE R3, RESULT.
```

If each of these instructions is 18 bits long, then we have used a total of 72 bits, plus whatever base register loading we do for the loads and stores. This is disappointing since the original three full address ADD was 60 bits. However—and this is why it is so difficult to make decisions in this area—examples can be found in large quantities to support the idea that a load/store form is more efficient or that a three-address form is more efficient: Consider

```
H=B*C;
G=D*E;
A=H-G;
```

and the following code:

$$
\begin{array}{lll}
\text{MULTIPLY} & \text{B, C, H} \\
\text{MULTIPLY} & \text{D, E, G} \\
\text{SUBTRACT} & \text{H, G, A}
\end{array}
$$

A problem with the storage-to-storage (no register) three full address architecture is that it has no register to hold the intermediate values, G and H. If the instructions are each 60 bits (6 bits of op code and 18 bits of address and index for each operand), then 180 bits are required to do the assignment statements. In addition, there is an additional space requirement for G and H and additional data bit flow as G and H are stored and retrieved. If the instructions are compressed with the use of a base register, then each might be 36 bits, requiring 108 bits to do the program computation. The need for storage and retrieval of G and H remains, however.

Consider the coding for a register file architecture in which all operand references for arithmetic were made in registers:

$$
\begin{array}{l}
\text{LOAD R1, B} \\
\text{LOAD R2, C} \\
\text{MPY R1, R2, R3} \\
\text{LOAD R1, D} \\
\text{LOAD R2, E} \\
\text{MPY R1, R2, R4} \\
\text{SUB R3, R4, R5} \\
\text{STOR R5, A}
\end{array}
$$

Now, of course, we are assuming that G and H were truly only intermediate values. We see that we have many more instructions to do. The total bit count, however, might be held to 144 bits. This assumes a base register and 4-bit register address giving an 18-bit instruction. This is better than the 60-bit word form of storage-to-storage. It is also better than the compressed 36-bit three-address form because it does away with the space required for intermediate variables G and H. The three-memory address machine could do no other than pick locations in memory for storing G and H. Thus we see two things: that there is a real convenience to registers and that one can pick a scenario to show that one approach to architecture is better than another with almost equal convenience.

## 4. FEWER ADDRESSES

Another way of shortening address space is to reduce the number of addresses that are provided in an instruction. Thus a result address may be implied as one of the operand addresses. This may lead to the following code for A=B+C on a register file architecture:

LOAD R1, B
LOAD R2, C
ADD R1, R2 (Result to R1)
STORE R1, A

An architecture using only two addresses leads us to a general argument among architects that is essentially whether A=B+A is a more common form than A=B+C. Clearly, if A=B+C is more used, then a three-address form, register, or memory address is more effective than a two-address form. If, however, A=B+A is more common, then the space required for the representation of the third address, whether of register length or memory length, is wasteful. For the comparatively few occasions of A=B+C, additional instructions are worthwhile since the bits for the third address will be useless most of the time. Some architectures confront this problem by providing two and three address forms, for both register and memory addressing.

As we observed when discussing stacks in the last chapter, some architectures undertake to eliminate the need for operand addresses in arithmetic instructions. Even a compiler for a stack machine, however, may face decisions about the best sequences of PUSHes and arithmetic operations.

## 5. MIXED ADDRESSING FORMS

Thus we see that there may be a very large number of approaches to addressing forms and combinations involving address shortening, implied addresses, and so on. We see in Chapter 12 that there is a lack of general agreement about the best form of an ideal architecture and that part of this disagreement concerns the form and number of addresses. There is a large set of complex and interrelated decisions to be made based on judgments about the forms of high-level language statements that will be used, the frequency of the forms, and so on.

When comparing the storage-to-storage three-address full memory location form of an instruction to the load register forms, we will see instances where it is convenient to intermix operand and register addresses in highly flexible ways.

In fact, many architectures use multiple addressing conventions and group instructions by the addressing conventions that are used. In some architectures the addressing forms are associated with particular operation codes and limitations on addressing are associated with the operation code. Thus an ADD instruction, for example, may use only register addressing, a LOAD instruction must reference memory, and so on. In other architectures there is some ability for an instruction to indicate what addressing forms it will use. This indication may be coded into the operation code so that there is the appearance of an ADD instruction that uses a register address for one operand and an indexed memory address for another, and an ADD REGISTER instruction that uses register addresses for both operands.

LOAD R1, B (32)      LOAD R1, B (32)

ADD R1, C (32)       LOAD R2, C (32)

STORE R1, A (32)     ADDR R1, R2 (16)

STORE R1, A (32)

The first observation that the additional load to support the ADDR form must be wasteful is modified by the consideration of the possibility that two loads will not always be necessary and that one or both of the operands will be in registers. We see when we discuss organization in Part Two that the additional instruction might not add to sequential performance time either.

In some architectures an operation code may specify how many addresses it will use, effectively indicating whether or not it assumes that the result will go to a unique place or to one of the operand positions. In some very flexible addressing schemes each address can specify what form it is, a full address, a partial address that needs a base, a register address, and so on.

The IBM 370 generic architecture uses multiple addressing forms associated with different types of instructions. In this architecture there are variable instruction sizes, the size of an instruction depending on the form of addressing used. There are

1. Register–memory instructions, which refer to one operand in a register and to another operand in a memory location.
2. Register–register instructions, which refer to both operands in a register.
3. Memory–memory instructions, which refer to both operands in memory.
4. Immediate instructions, which contain their operand in the instruction.

In creating code for this architecture a compiler must make choices about what coding and address pattern to use. In effect the architecture provides limited instances of the accumulator, register file, and storage-to-storage architectures.

### 6.   REGISTERS IN ADDRESS FORMATION

Registers may be referenced in an instruction so that values in the registers may be used to participate in the formation of an address. When registers are so used they are used in a number of basic ways:

1. As index registers to add an offset value to select a particular element from a vector or array by combining the value in an index register with a base address occurring in the instruction.
2. As *base, displacement, or segment* registers, all of which carry the connotation that the base address is in the register and an offset value is in the

instruction. Thus the base, displacement, or segment register defines the range of addressing for a program by establishing a lower bound beneath which the program cannot address. It also establishes an upper bound for architectures that do not address all of memory in an instruction. The range of addressability is the range defined by the value in the base register plus the maximum value that can occur in the instruction address portion (plus any indexing values). Thus if there is a base register of 24 bits in an architecture whose memory is completely addressable by 24 bits, and 12 bits of address in an instruction, any program can refer to 4K locations in memory. The particular 4K locations are determined by the value in the base register. In this way some notions of protection coincide with some notions of address compression because a program cannot address below a base register set at the value 64K (K=1024). Its addressing range is 64K to 64K+4K. In fact there are some holes in this notion. Protection of memory beyond the range defined by the base register is dependable only if the program cannot itself change the base register and only if the address size in the instruction cannot address all of memory.

Some architectures protect the base and segment registers by restricting the base register modification instructions to control state. Others allow base registers to be defined within the general register population and freely modified so that they rely on some other mechanism for protection or provide no protection at all.

In architectures where the full memory is addressable from an instruction, an additional *limits* register may be used to define the range of addresses permitted to a program.

3. A register can be used to hold a full address. An instruction then specifies the register holding the address of an operand it uses. The register holding the address may be a special address register, a set of address registers, or a part of the general register set. A convenience for taking addresses from the general register set is the ability to load a register with an address and execute a BRANCH REGISTER designating the register with the address. There is also an ability to compute an address in a register that can be used for a branching or for any operand. In architectures that have some very short instructions, 8 bits long, memory addresses must be gotten from an address register loaded with the address by a previous instruction. Instructions take the form, for example, of LOAD R1 compressed into 8 bits where R1 means that the address is to be found in a (usually) 16-bit register and an accumulator operand model is implied.

## 7. INDEXING

The concept of *indexing* is so basic to computer architecture that we have been forced to refer to it already although we have not yet discussed it in any detail.

We have relied on a rather terse introduction to the notion and to the intuitive appreciation that a high-level programmer, used to forms like A(I), would have for the notion of modifying an address with a subscript value. We now describe indexing a little more fully.

Indexing characteristically provides an offset value in a register to be applied to a base value in an instruction word.

Consider the DO-loop:

$$\text{DO I=1, 10, 1;}$$
$$\text{A(I)=B(I)*C(I);}$$
$$\text{END;}$$

On a processor without indexing capability it would be necessary to either code *straight-line* to achieve the intent of the loop or to modify every instruction after every iteration. Obviously neither is a desirable programming convention. The straight-line approach requires additional space since each iteration would be represented by unique coding. Instruction modification is undesirable from the point of view of time required to modify instructions and because contemporary program methodology abhors instruction modification because of the debugging difficulties that may come from it. As we mentioned in Chapter 3, some architectures do not permit instruction modification because a tag identifies a location as an instruction and prohibits that location from being modified.

In order to avoid straight-line code or instruction address modification, an index register value is added or subtracted from a base address in the instruction word. This value is changed for every iteration of the loop.

The indexing concept is fundamental to almost all architectures. In its basic form there is a single index register that can be loaded with an initial value representing the relative position of an element in an array. A bit in an instruction word shows whether or not the indexing value must be added (or subtracted) from the address in the instruction word. At the end of the loop a specialized index register manipulation instruction (e.g., TXI—Test Index and Increment) modifies the index register value and tests it for a value. If the value in the index register is the limit value, then the next instruction to be executed is the instruction following the index test instruction, if not, the next instruction is the beginning of another iteration of the loop. Exit from the loop occurs when the test instruction finds the index register at the limit value. A single index register architecture will use the index register to modify operand addresses and DO-loop end control. For example:

$$\text{DO I=1, 10, 1;}$$
$$\text{A(I)=B(I)*C(I);}$$
$$\text{END:}$$

```
          LOAD INDEX ("0")
LOOP:     LOAD R1, B, *
          MPY R1, C, *
          STORE R1, A, *
          TEST INDEX ("10"), LOOP
```

The above code loads the index register with an immediate value of 0. It then loads register 1 with the element of the array B that exists in location B + (INDEX). The element from array C is referenced as location C + (INDEX), and the result array, A, is referenced as A + (INDEX). This assembly language uses the convention that values in " are immediate values, and the index register reference is indicated by *.

The first iteration of the loop will be executed with the value of 0 in the index register. When a result is stored in location A + 0, the TEST INDEX instruction will increase the index register with a value of 1, setting it equal to 1. The instruction will then test to see if the value is equal to 10. If the value is not equal to 10, then a branch to label LOOP will be executed. Another iteration of the loop will be performed with array elements B+1, C+1, and A+1 referenced by the code. After 10 iterations, the TEST INDEX instruction will add a 1 to the index value that will make it equal 10. At this time no transfer back to LOOP will occur and the instruction following TEST INDEX will be executed; the loop will have been completed.

The above example is only one way of using an index register to both modify addresses and serve as a counter. The architecture may have an instruction that tests before it adds. That instruction might be put at the beginning of the loop. In some architectures it is possible to use a single index register to form addresses but a decrement memory instruction as a counter. Multiple index register machines may use one index register to form addresses by adding and another as a counter that is decreased to 0 for the limit test.

In order to avoid counting up in an index register and providing a constant for limit testing, some compilers have placed arrays in memory backwards. For example, a five-element array would be arranged in memory:

| Element | Location |
| --- | --- |
| A (1) | 1004 |
| A (2) | 1003 |
| A (3) | 1002 |
| A (4) | 1001 |
| A (5) | 1000 |

This permits the derivation of count and address by counting down in the index register until it reaches 0. In architectures that allow only subtraction this is an important approach. In any case, it eliminates the need for a constant to establish the initial value of the index register and another constant to test against.

There are several extensions and variations on the indexing theme. An interesting feature of UNIVAC 1100 architecture is the provision of a decrement or increment value in the upper portion of an index register itself. Thus in a 36-bit UNIVAC 1100 index register, there is an 18-bit value, a 16-bit increment or decrement, and a 2-bit field that indicates whether the upper value is an increment or decrement. An instruction on the UNIVAC 1100 may call for an index modify function that applies the increment/decrement field to the index register value to be performed.

| Code | Increment/Decrement | Index Value |
|------|---------------------|-------------|
| DEC  | 5                   | 1000        |

After execution of an instruction with an *index modifier* bit set on:

| Code | Increment/Decrement | Index Value |
|------|---------------------|-------------|
| DEC  | 5                   | 995         |

Thus the U1100 architecture automatically modifies index register values as a suboperation when specified in an instruction using an index register.

Most processors provide more than one index register so that the designation of indexing in the instruction word gives the register designation of a particular index register rather than just specifying that indexing is required.

Multiple index registers allow for the addressing of multidimensional arrays. A reference, for example to A(I,J,K), may conveniently use an index register for each dimensional offset. In some architectures an instruction may show that it wishes three index registers to be involved in address formation for a particular element. Thus, on a machine with three index registers the binary value 000 in the index register designation field may indicate no indexing; 001, the use of index register 1; 010, the index register 2; 100, the index register 3. Beyond this:

| | |
|-----|-------------------------------|
| 011 | Use index registers 1 and 2   |
| 101 | Use index registers 1 and 3   |
| 110 | Use index registers 2 and 3   |
| 111 | Use index registers 1, 2, and 3 |

If the operands of an instruction can have different sizes, the increment or decrement values applied to the index register that is used to form an address must reflect this fact. In the VAX-11, adding to or subtracting from an index register is automatically adjusted as a result of the coding of an instruction that indicates whether byte, word, longword, or quadword elements are being used. Thus an array of 4-byte elements is automatically indexed by adding 4, an array of 2-byte elements by adding 2, and the modification to the index register is determined by element byte lengths.

## 8. PAGE, BASE, AND SEGMENT ADDRESSING

Values in index registers are offsets to addresses provided in the instruction. Values in base type registers provide a base address, and the value in the instruction word is considered an offset to this base address.

A memory address in an instruction may be a full or partial address. The number of bits permitted in the instruction for the representation of an address may not allow for reference to be made to all of memory. A full memory address is formed by the addition of the address in the instruction to the contents of a register, or the concatenation of the address in the instruction with a value in a register.

When the address in the instruction is concatenated with a value in a register, the address form is sometimes called *banking* or *page selection*. The value in the page register represents the upper bits of a full address and the value in the instruction the lower bits. If, for example, there are only 8 bits of address allowed in an instruction, the maximum address that can be formed is 255. In order to address beyond a 256-word area, a value is placed in the page register that provides for wider addressing. For the class of instruction using this addressing form, the register contents are always appended in front of the address in the instruction. The bits of the value in the register are interpreted as higher-order bits of, say, a 16-bit address. An 8-bit register would provide for addressability to 32K addresses. A form of addressing called *page zero* addressing provides for the addressing of the low 256 locations of memory by providing suppression of the concatenation of the contents of the page register.

Generalizations of page relative addressing provide for a choice of registers to specify as base registers and for the addition, rather than the concatenation, of the values in the specified register.

The table below shows a 12-bit address, taken from an instruction, added to a 24-bit address in a base register to form a full address. The address in the instruction word is a displacement indicating a relative position in the area of memory whose starting position is indicated by the value in the base register.

000000001100000000000000 + 000001101010
Base register: 49152          Address field: 0106
FINAL ADDRESS: 49258

Some architectures allow referencing relative to the value represented in the instruction counter. In effect the instruction counter can be used as a base register. A common assembly language form is the following:

1000 LOAD *+210
1001 ADD *+210
1002 STORE *+210
1003 JUMP *+100

This coding indicates that values at 1210, 1211, and 1212 are to be used and that the next instruction will come from location 1103. Such addressing is called *location relative addressing*, and the JUMP instruction is called a *relative jump*. References to location minus value are usually permitted as well as references to location plus value. The form provides a limited addressing range with a constricted relative value.

As we mentioned in the introductory section, some architectures treat base registers as control registers and constrain the ability of applications programs to modify them. In these architectures a major function of protection is seen for the base registers. Other architectures allow free addressing and modification of base registers, either as sets of special registers or within the general register population. If more than one register may be used as a base register, then different parts of a program may be *covered* (relative to) different base registers. This leads to some concepts of program structure and program partitioning.

## 9. SEGMENTATION

Segmentation is a word with a diversity of meanings in different architectures. Segmentation is essentially a form of base addressing. Some architectures use the word segmentation to describe what others call base addressing; some architectures describe segmentation as something in addition to base addressing. The notion of segmentation is that a program, or a set of programs, may be organized so that a set of base addresses is convenient. Within the context of a single program, various parts of the program may have different base locations.

The concept of segmentation appears in two basic ways in an architecture. Specific function segmentation allocates a base (or segment) register to very particular parts of a program structure. There may be one segment register used to reference coding, another to reference data, another to refer to specialized data working area, and so on. The INTEL 8086, for example, has four segment registers of 16 bits, each of which is associated with a particular portion of memory as described above.

Full addressing of an INTEL 8086 memory requires the formation of a 20-bit address. Values formed by index, base, and instruction displacement arithmetic, however, are 16 bits long so that full addressability to memory cannot occur using just base register, index register, and instruction address arithmetic. The segment registers of the machine are also 16 bits long, but they are assumed to be 16 high-order bits with 4 low-order bits of 0. The addition of a 16-bit address and a segment register value creates a 20-bit address.

The use of segment registers in this machine provides another level of address formation beyond the index and base register value. A segment is a 64K structure that can start in any location divisible by 16. The exact location in a segment can be referenced by a 16-bit instruction address or a 16-bit instruction address as modified by an index register, a base register, or the sum of index and base registers.

Instruction sequencing is provided through a 16-bit instruction counter and a *code segment register*. Operand addressing is provided through a *data segment register* that holds the starting address of an area of memory used to hold data. If no indication is given otherwise, the data segment register will be used to form a 20-bit operand address after base and index register operations are performed. (There is a slight simplification in this description.) An instruction may have a 1-byte prefix that provides the address of the specific segment register that is to be used in forming the address.

Thus the 8086 segmentation does two things. It provides full addressability, needed because of the history of the architecture, and it provides a structuring concept for programs into data, program, and stack segments.

Another concept of segmentation is used with the UNIVAC 1100 architecture. In this architecture there is no implied special use of a particular segment register. A hardware algorithm is used to determine which segment register should be used to form a full address. This hardware algorithm is needed because there is no room in a UNIVAC 1100 instruction word to indicate which segment is desired.

In the UNIVAC 1100 architecture the segmentation registers provide the function of base registers. The additional level of address formation that exists with the INTEL 8086 does not exist here.

The UNIVAC 1100 architecture contains four segment registers that provide base addresses for four segments of memory. A segment is a space of arbitrary size. Each segment register holds a segment descriptor entry that provides a base address for a memory area (the size in bits of this base is model dependent) and upper and lower limits for the area. A relative address, formed from the instruction word and an index register, is added to the base value in a selected segment register to form a full memory address.

The exact memory formation algorithm is model dependent in the UNIVAC 1100 architecture, but in general the algorithm works as follows: At any one time a particular set of segment registers is considered to be active. A segment register set consists of two segment registers. When a set is active, one of the two segment descriptors associated with the set will be used to form an address. The hardware tests the relative address against the limits of the first segment descriptor, and, if it is not exceeded, the base of the first descriptor will be used to form a full address. If the bounds are exceeded, the second segment descriptor will be used.

## 10.  INDIRECT ADDRESSING

An additional addressing convention associated with some architectures is *indirect addressing*. An indirect address is interpreted to mean an address that is not the address of the operand required but an address that holds the address of that operand. The instruction word contains a bit that shows whether or not indirect addressing is to be used. The usual formation of a full address, using base and/or index registers, is undertaken by the processor, but when the address is formed its contents are subjected to an additional cycle of address formation which leads, finally, to the address of the desired operand. In some architectures indirect addressing can be *cascading*. This means that an *indirect address word* (a word pointed to by an indirect address) can contain a bit indicating that its contents represent the address of another address. The process of address formation continues until an address word is found that does not have its *indirect bit* on.

The major purpose of indirect addressing is to provide relocatability for referenced data. An operating system may find it convenient to determine locations for programs and for data referred to independently. The relocation of data elements without disturbing the addresses in a program may be accomplished by routing addresses for relocatable elements through an indirect address table. The movement of the elements will cause changes in the indirect address table, but the values of the addresses in the program may remain invariant.

Some architects and designers feel that the development of virtual addressing concepts (see Chapter 10) eliminates the need for indirect addressing. Others feel that indirect addressing is inefficient since it requires multiple references to memory for an operand and considerably enlarges the bit flow for the execution of functions. As we see in Chapter 12, one theoretical criterion for a good architecture is that only one reference to memory is made for each operand.

### 11.  MEMORY PROTECTION

We have seen that addressing techniques and memory protection are closely related. In a base register architecture with short addresses in instructions only the range defined by the base register and the partial address in the instruction word can be referenced. To provide protection, the architecture can constrain the ability of a program to change base register values. It can defeat an attempt to develop an address, by indexing or some other form of address arithmetic, beyond the limits by the generation of a *protection exception*. The architectures that rely on the addressing mechanisms to achieve protection rely on an operating system running in control state to place values in base or segment registers.

Some architectures, however, use protection schemes that are not identical with the addressing schemes. Protection is perceived as an issue that is separate from basic address formation. This may be true of an architecture for any number of reasons:

1.  The addressing scheme is "open." That is, the registers holding base values may be freely manipulated by a running program that is not in control state. As a consequence, the base registers cannot be relied on to constrain addressing.

2.  The desired *granularity* of protection is finer than the granularity of addressing provided. By granularity we mean the number of addresses that can be protected by the same parameter or value used in the protection mechanism.

    Consider a base address of 16 bits and an instruction address of 12 bits. When the base address is set with a value, a program can address 4K locations from the base. It may be desirable, however, not to allow a program to address every location in that range. If it is an area shared by programs or by a program and the operating system, we may wish to designate certain nonpermissible locations within the addressing range and consequently need an additional protection mechanism.

3.   Various types of addressing may be permitted and others excluded. For example, it may be permissible to read (take a value from) a location in an addressing range but not change it, to take an instruction from a location but not use the location for an operand. Thus some extension to the basic addressing scheme or some auxiliary mechanism is required to denote only partially addressable locations.

The basic protection mechanisms that we describe in this chapter and that have meaning outside of the concept of virtual memory are *protection keys* and *protection rings*.

## 11.1.   Protection Keys

A protection key is a value that can be assigned to a group of memory locations. A program trying to address these memory locations must be able not only to develop the address but also to present the key. The key is associated with a running program by the operating system and is kept in a control register. A set of special, non-program addressable memory locations holds the keys associated with a particular area. The granularity of an area protected with a particular key may differ from architecture to architecture and from machine model to machine model of an architecture. A reference to a memory area with a key that is different from the program key will cause an addressing alert in the processor.

Some architectures permit a key that allows addressability to all locations in the machine. Operating systems code is frequently allowed to operate with this protection key so that the operating system can address through the system freely. The idea of protection is to protect running programs in a set of sharing programs from each other and to protect the operating system. Some programming systems specialists do not like the idea of the operating system running with no constraints for security reasons.

## 11.2.   Protection Rings

The idea of protection rings first appeared in operating systems design and is associated particularly with the MULTICS operating system. Since the introduction of the concept a number of hardware architects have undertaken to design hardware-oriented facilities to support the concept.

A ring of protection involves a hierarchic program structure in which programs are visualized operating at different levels of privilege. A common example is the execution of a set of programs written by students. These programs share the machine with a checking and correcting program written by the professor. The programs of the students should not be able to access the answers and cannot be allowed addressability to the professor's program. The professor's program, however, does not have complete control over the machine but runs with the assistance of an operating system. This operating system may be divided into a *services section* and a *kernel*. The services section provides certain functions for the program, I/O coordination, for example, and the professor's program is permitted to ask for execution of the services program. However, it is not permitted to address the tables and values that the services programs use for service functions. The services programs use functions provided by the kernel. The kernel performs basic machine functions and is in control of setting base registers and protection ring values, for example, or of using other mechanisms for controlling programs above it.

A usual visual representation of this notion is as follows:

LAYER One       STUDENTS' PROGRAMS
LAYER Two       PROFESSOR'S PROGRAM
LAYER Three       SERVICES OF OPERATING SYSTEM
LAYER Four       KERNEL

We associate each of the layers of the hierarchy with a ring. Thus we say that the students' programs lie in ring 1, the professor's program in ring 2, and so on. In some architectures the "name" or designation of a ring is identical to a segment address: in other architectures a segment address and a ring designation are different and individually represented.

When ring designations and addressing are separate it is possible for a program to develop an address that constitutes a reference to another ring. This is known as a *ring crossing*. Hardware will recognize the attempt to cross a ring and determine whether the ring crossing is permissible. If it isn't, a ring protection alert will be generated.

The association of ring identification across a set of segments or other addressing spans may be quite arbitrary and rings of protection consequently of different size. A running program is associated with a ring and can address other programs and data within the ring regardless of the segment reference patterns that are involved. Part of the operating system designation of the privilege of a program involves the definition of what rings it may refer to. A program may commonly cross rings at and above its own level, but it is restricted from addressing in rings below it. Thus a hierarchy of privilege is defined by the ring structure.

We speak of a programming as running in a certain ring, meaning that it runs at a certain level of privilege. One feature that may be associated with rings is the selection of a different ring level for the same program, depending on whose behalf the program is running. Thus a program derives its ring designation for a particular execution as a result of the rights of its programmer or user.

## QUESTIONS

1. What is the use of indexing?

2. Describe base displacement addressing. Why might it be desirable to constrain the addressing range of addresses in an instruction?

3. What is page relative addressing?

4. Describe indirect addressing.

5. What is the use of segmentation on the INTEL 8086?

6. Why do architects desire to reduce the number of bits required to address operands?

7. Why do you think the CRAY-1 does not allow arithmetic instructions to address memory?

8. Why might a protection mechanism beyond the addressing conventions be necessary?

9. What is a ring?

# Chapter 6

# Instruction Sets

## 1. INSTRUCTION SET ORGANIZATION

The instruction set of a processor is the list of fundamental operations that the machine can perform on data or on elements of its own control mechanisms. Instruction sets differ in many ways: in the set of operation codes provided, in the variety of ways that operands may be referenced, in the methods used to code operation codes, in the ability of various operations to use various operand addressing forms, and so on. These differences are partially due to various views of architectural efficiency and partially due to some concept of the primary use of the computer. Although the instruction sets of most computers are extensive enough to allow the computer to be used for a broad class of problems, as a general-purpose computer, some architectures are biased toward a particular efficiency for a particular type of application. Thus we have architectures that are most efficient for high-speed scientific computing, for network interfacing, and so on. This diversity in architectures forces us to choose among architectures and creates some problems in comparing the efficiency of one machine to another.

In general, an instruction contains an operation code and one or more operand addresses. In addition, there may be some special fields that modify an operation in some way by providing additional information about operand handling or address interpretation.

## 2. INFORMAL GROUPINGS OF INSTRUCTIONS

The specific classification of instructions that form an architecture depends very much on the architecture itself. It is possible to partition the instructions of an architecture into different groups of instructions in a variety of ways. One informal general grouping of instructions is as follows:

1. *Data Movement Instructions.* These instructions move the contents of memory locations between processor registers and memory, or from one mem-

74

ory location to another. They do not change the data that are moved. Such instructions are LOAD, STORE, and MOVE.

2. *Arithmetic Instructions.* These instructions perform arithmetic operations on data, commonly ADD, SUBTRACT, MULTIPLY, and DIVIDE. There may be special instructions for different data representations giving rise to ADD, ADD DECIMAL, FLOATING ADD, and so on.

3. *Logical Instructions.* These instructions perform manipulations on data in terms of the basic boolean operations of OR, AND, NOR, NAND, EXCLUSIVE OR, and EXCLUSIVE AND. The data are treated as streams of bits, and corresponding bit positions are processed against each other in accordance with the logical rule being applied. The following tables represent the results of applying various logical operators to various patterns of bit values to which a logical rule is applied. We show OR, AND, NOR, and NAND:

OR

| Bit Value | | |
|---|---|---|
| A | B | Result |
| 0 | 0 | 0 |
| 0 | 1 | 1 |
| 1 | 0 | 1 |
| 1 | 1 | 1 |

*Rule:* Output will have a 1 bit if a 1 bit occurs in a corresponding position in either input.

NOR

| Bit Value | | |
|---|---|---|
| A | B | Result |
| 0 | 0 | 1 |
| 0 | 1 | 0 |
| 1 | 0 | 0 |
| 1 | 1 | 0 |

*Rule:* Output will have a 1 bit if no 1 bit occurs in a corresponding position in either input.

AND

| Bit Value | | |
|---|---|---|
| A | B | Result |
| 0 | 0 | 0 |
| 0 | 1 | 0 |
| 1 | 0 | 0 |
| 1 | 1 | 1 |

*Rule:* Output will have a 1 bit if a 1 bit occurs in corresponding positions in both inputs.

NAND

| Bit Value | | |
|---|---|---|
| A | B | Result |
| 0 | 0 | 1 |
| 0 | 1 | 1 |
| 1 | 0 | 1 |
| 1 | 1 | 0 |

*Rule:* Output will have a 1 bit if a 1 bit does not occur in corresponding positions of both inputs.

These boolean operations are used in a variety of situations. They are used to edit data when no other editing facility exists in the machine. If one of the inputs is a preset pattern of bits called a *mask* and the other input is a data field, it is possible to delete 0's and perform other transformations using the logical operations.

The boolean operations may be used in situations where the outside world is sending data in such a way that the presence or absence of a certain condition is represented in master bit fashion in a single computer word. For example, if the processor is monitoring a set of on-line devices the status of each device, on or off, may be represented in a single bit position. When the processor wishes to discover the state of a particular device it performs a boolean operation on the word, zeroing out status bits for all other devices and isolating the status bit for the device under investigation.

| D1 | D2 | D3 | D4 | D5 | D6 | D7 | D8 |
|---|---|---|---|---|---|---|---|
| 0 | 0 | 0 | 0 | 1 | 0 | 1 | 0 |

The above indicates that devices 5 and 7 desire to communicate with the processor. In order to determine that device 7 is active, the 8-bit status word shown above can be combined in an AND with the mask 00000010. The result of this AND is to isolate the value 00000010 in a register. A test on register value can then determine the activity of device 7.

Somewhat later we will see that the boolean functions OR, AND, NOR, and NAND are interesting because electronic circuits representing these fundamental conditions are the basic building blocks of computers. They are used in various combinations to represent the conditions under which signals and data flow from one electronic gate to another in the design of computer logic.

4. *Editing Instructions.* These instructions operate on character strings as an extended MOVE with associated logical operations. They suppress 0's and insert punctuation into a character string as it moves from a source to a target set of memory locations.

5. *Shifting Instructions.* These provide the ability to move the bit contents of a word a designated number of positions to the left or to the right. They are used for editing and for field alignment before or after arithmetic operations to effect rounding. There are commonly SHIFT LEFT and SHIFT RIGHT operations of various kinds. The instruction designates the number of bit positions to shift the contents of a processor register. Variations include the ability to shift across register boundaries, to shift without disturbing the sign position of a word, to count the number of shifts until a 0 bit is encountered, and to end around shift and move the upper bit positions into low-bit positions in the word.

6. *Immediate Instructions.* These provide for the representation of values rather than addresses in the instruction word. An instruction like ADD IMMEDIATE shows that the specific value represented in the address field is to be added to another operand. Some architectures do not distinguish between immediate and other instructions in the operation code but permit an ADD instruction, for example, to refer to operands using any number of addressing forms, among them the coding of the operands in the instruction in place of an address.

7. *Index Register Manipulation Instructions.* These instructions load, modify, and test the contents of a designated index register. They are intended to provide address modification and loop control. An architecture may have instructions that cause the modification of an index register value and a test for 0 or against a specified limit.

8. *Branch Instructions.* These provide a way to alter the flow of instructions unconditionally or on the determination of a condition of data. They provide a means of testing for a condition, such as a 0 value in a register or that the contents of one value is equal to another, and they provide an address for the next instruction to be executed if the stated condition is true. A computer architecture may have several means of altering the flow of a program. These may be simple branch instructions or more complex instructions where branching is associated with another function. For example, there may be a CALL or RETURN function that is conditional on the existence of a certain value in a

register. Aspects of branching are discussed in Chapter 7, of subroutine linkage in Chapter 8.

9. *Memory Modification Instructions.* These instructions modify a value in memory by performing an arithmetic or logical function and returning the result to an indicated memory location. Instructions may add or subtract the value of a register to a particular memory location, increment or decrement a location with a value provided in the instruction, or add an implied value to a memory location. These instructions enable the efficient use of memory locations as counters, control variables, and so on. In many architectures these do not exist as a class of instruction but are made available by using an operation code that can use arbitrarily selected addressing forms.

10. *Formal Subroutine Linkage Instructions.* Some architectures have instructions in the hardware that are intended to increase the efficiency of subroutine and procedure invocations. Thus a single CALL instruction may place the return point of a linkage in a specific location in the processor, transfer control to the invoked subroutine, and pass parameters in a formalized way without a set of supporting instructions. A LIFO stack is a frequent mechanism for supporting the functions of subroutine linkage and procedure invocation.

11. *Stack Manipulation Instructions.* Some architectures have a concept of a LIFO stack of registers or memory locations. Stack manipulation instructions are either implicit or explicit. An implicit instruction changes the stack by referencing the stack in the context of another operation, such as ADD. On a stack machine an ADD may be interpreted to mean add the two values at the top of the stack to each other and place the result on top of the stack, removing the two operands. Explicit instructions call for the direct manipulation of the stack. An example is POP, take a value off the stack, or PUSH, make room for another value on the stack. Operand stacks defined as sets of operand registers are discussed in Chapter 4, stacks defined in memory for use as linkage stacks are discussed in Chapter 8.

12. *I/O Instructions.* These direct the flow of data to and from the devices that surround the processor and its memory. They provide the address in memory of where data is to be put (or from where it is to be taken), the amount of data that is to be moved, and the address of the device that is to receive or provide data. Device addressing conventions differ from memory addressing conventions in most machines.

Sometimes the I/O instruction itself only provides the address of a set of locations in memory where the information needed to control an I/O operation may be found.

13. *Privileged or Control Instructions.* These are instructions that can only be executed in *control state*. Control state is a machine status entered by the execution of an instruction or by the occurrence of an interrupt. When in control state certain instructions may be executed that affect the memory addresses an application program may reference or that change the contents of certain control registers. These instructions are commonly executed by the

*operating system control program* that is responsible for the coordination of applications programs that share the computer. The intent of the privileged instructions is to prevent an application program from undertaking an action that might cause an erroneous condition for another program. LOAD BASE REGISTER, on a machine where base addressing is used (see Chapter 5) may be a control instruction.

### 3. OTHER INSTRUCTION GROUPINGS

There are various ways to group instructions other than as shown above. Specific instruction groupings are always influenced by details of a particular architecture. Instructions may be grouped by various criteria:

1. By the addressing conventions we have seen in Chapter 5, some architectures group instructions in R–R form, both operands in registers; R–M form, one operand in register, another in memory; M–M form, both operands in memory; and Immediate, operand(s) imbedded in instruction.

2. In some architectures memory referencing characteristics lead to instructions of different length and a consequent grouping of instruction by size.

3. By size of the operand referred to, leading to characterizations of fullword, halfword, byte, or block instructions.

4. By the data representation of operands on which they operate. Thus a system may have a set of binary instructions, a set of floating point instructions, a set of decimal and character manipulation instructions, a set of vector manipulation instructions, and so on.

### 4. SYMMETRIC INSTRUCTION SETS

Some workers refer to a concept of the *regularity* or *symmetricity* of an instruction set. A perfectly symmetric instruction set contains identical functions for all data representations and for all addressing modes. The instruction set has a set of basic operation codes that are associated with coded modifiers to determine addressing and data modes.

An example of a highly symmetric instruction set is found in the architecture of the Digital Equipment Corporation's VAX-11. In this architecture there is a set of basic instructions for integer manipulation that consists of MOVE, ADD, SUBTRACT, MULTIPLY, and DIVIDE instructions, among others. Operands in the VAX-11 may be 1 byte long (byte form), 2 bytes long (word form), 4 bytes long (longword form), or 8 bytes long (quadword form). An instruction (shown in assembly language form) to move various operand lengths would be written as follows:

| | | |
|---|---|---|
| MOVB | OPERAND1, OPERAND2 | MOVE ONE BYTE |
| MOVW | OPERAND1, OPERAND2 | MOVE ONE WORD |
| MOVL | OPERAND1, OPERAND2 | MOVE ONE LONGWORD |
| MOVQ | OPERAND1, OPERAND2 | MOVE ONE QUADWORD |
| MOVF | OPERAND1, OPERAND2 | MOVE FLOATING POINT WORD |

Each of the above instructions is coded in a 1-byte instruction field. The length of a VAX-11 instruction is determined by the addressing conventions used with the instruction. Many instructions may have both two and three operand forms. The 1 byte containing the operation code will indicate whether a two- or three-operand form is being used. In two-operand form the two operands are operated on and if a third operand is needed (e.g., the place to put a result) the location of the third is implied to be the location specified by the second operand. In three-operand form a location is provided for the result.

The actual size of a VAX-11 instruction depends not only on how many operands are specified, but on the forms of the addresses used for each operand. The basic size of a two-operand instruction word is 3 bytes. One byte for operation code (including the designation of data unit size and coding), and one byte for each operand address. Part of the byte for each operand address contains an indicator of the type of address being used. When operands are in registers, all addresses are 1 byte long. The register address is in the lower portion of the byte. The upper portion defines the byte as a register address. Certain addressing modes require additional bytes to provide an operand address. For example, an address of an operand in memory will require multiple bytes. Each address type is self-described in the upper bits of the first byte. In effect, the addressing schemes provide the ability for an instruction to reference operands in registers, operands whose addresses are in registers, and operands that exist in memory. The addressing scheme is independent of the operation code. The maximum size for a VAX-11 instruction is rather long. In general a complete operand address can require 9 bytes, and some instructions can reference six operands. Thus the VAX-11 architecture is highly symmetric as regards the location of operands as well as the size and coding of an operand.

There are architectures that do not have instructions for identical operations on all data types, or that restrict the addressing conventions that may be used with particular instructions. In the Control Data Corporation CYBER architecture, for example, arithmetic instructions cannot use memory addresses, but must depend on the operands being in registers at the time the arithmetic instruction is executed. In the generic 370 architecture there is provision for memory-to-memory arithmetic only for packed decimal data.

An important aspect of a regular instruction set is that instruction side effects be regular. Some machines set indicators, called *condition codes*, to represent attributes of the data after the execution of an instruction. For example, the data may be represented as being greater than, less than, or equal to 0 after being moved into a register or used in an arithmetic function. In a regular architecture these condition code setting conventions would be uniform regardless of the coding of the data. In an irregular instruction set, condition codes might be set for binary or decimal data but not for data moved into a register by character or string manipulating instructions. As another example, a symmetric architecture would have a convention for register zeroing that applied whenever a memory location holding a value smaller than the register was moved into a register. An irregular architecture might use different conventions, depending on whether the field was moved by a character movement instruction or an arithmetic load instruction. A perfectly symmetric instruction set generates carries as a result of arithmetic in the same manner regardless of the bit length of the value being operated on.

The importance of regularity or symmetry to some architects is that the creation of compilers is made much simpler when side effects are uniform for all instructions, when all addressing forms are available for all instructions, and when all instructions operate on all data forms. The number of "special cases" that a compiler must look for to produce appropriate code is reduced. In addition, subtle bugs are eliminated at the machine language level.

## 5.  EXAMPLE INSTRUCTION FORMS

Some example formats of instruction are shown below. We have picked the UNIVAC 1100 as an example of an architecture with a fixed instruction size that uses one word per instruction. The Data General NOVA is presented as an architecture with a shorter instruction word that uses multiple instruction formats. Some comments about the INTEL 8080 and the Cray Research CRAY-1 conclude our discussion. Each of these architectures represents a different type of computer, although our notions of computer types are rather informal. The UNIVAC 1100 is a large-scale, general-purpose computer. The Data General NOVA is a typical mid-sized minicomputer. The INTEL 8080 is a microprocessor. The CRAY-1 is a high-performance large scale processor with a strong orientation toward scientific calculations involving significant amounts of floating point computation on vectors and arrays.

### 5.1.  UNIVAC 1100

This example format comes from the 36-bit word instruction of the UNIVAC 1100 architecture. The general form is as follows (top line gives characterization explained below, second line gives number of bits):

| OP | FS | AR | IR | MOD | IND | ADDRESS |
|----|----|----|----|-----|-----|---------|
| 6  | 4  | 4  | 4  | 1   | 1   | 16      |

The reader may wish to refer to the description of the U1100 operand register model provided in Chapter 4. The instruction set is a fixed word, register-to-memory set where:

1. OP is OPERATION CODE, which specifies the instruction to be performed. Actually there is a class of instructions where there is a 10-bit operation code and the FS field, discussed below, is used as an extension of the operation code. Thus the instruction set is really a two-format set. The coding of the first 6 bits of operation code indicates whether a 6-bit or 10-bit operation code is being used.

2. FS is FIELD SELECT, which may designate that only certain bit positions are to be transferred between operand registers and memory. The value of FIELD SELECT bits may call for selected groups of 6, 9, 12, or 18-bit fields to be selected from a 36-bit word. On a register load these values are brought to a register and right justified (placed in low-order bit positions).

3. AR is ARITH REGISTER, which specifies the arithmetic register to be used in the operation. The 4-bit field allows for a programmer to choose 1 of 16 for operand manipulation.

4. IR is INDEX REGISTER, which specifies an index register to be used in forming an address. One of 16 registers may be designated as an index register. There is partial operand and index register intersection.

5. MOD is INDEX MODIFY, which specifies that a value in the upper portion of an index register should be added or subtracted (as indicated in the upper 2 bits of the index register) to the lower 18 bits of the index register. This feature of the UNIVAC 1100 architecture provides for particularly efficient loop execution. In place of the use of a special index register modify instruction that adds or subtracts a value to an index register, index register modification can be called for as part of an arithmetic operation.

6. IND represents INDIRECT ADDRESS, which indicates that the address formed by the application of the index register value to the address field of the instruction should not be interpreted as the address of the desired operand. Rather, it is an address where the address of the operand may be found. An indirect address word referenced in this way may itself have an IND bit on, indicating that indirection is to continue.

7. ADDRESS specifies the location of an operand if no indexing is used. It specifies a value to which index register values should be applied if indexing is used.

## 5.2. Data General NOVA

This is a typical 16-bit instruction, taken from the format of the Data General NOVA architecture:

| FMS | OP | AR | IND | IR | ADDRESS |
|-----|----|----|-----|----|---------|
| 1   | 2  | 2  | 1   | 2  | 8       |

1.  The FORMAT SELECT FIELD (FMS) suggests that the architecture has multiple instruction formats. The interpretation of specific bit positions must be made in the context of the instruction format being used.

    Format selection is common to many architectures. It is indicated by the specific coding of an operation code. Operation coding may be such that all instructions whose first 2 bits have a 00 value use one format, those that have 01 use another, and so on. Some architectures do not explicitly express the concept of format selection; the U1100 and 370 are two of these. The format of an instruction is implied by the total numeric value of the operation code. Some machines, like NOVA, distinguish between format selection values and the operation code. In the NOVA, the four instruction formats are Arithmetic and Logic, I/O, Move Data, and Jump and Modify Memory.

    The NOVA format used is always partially suggested by the first bit of the instruction word. If the bit is 1, then the arithmetic and logic instruction format is used. If the bit is 0, then the second and third bits of the instruction indicate what format is being used. Initial 3-bit values of 011 indicate an I/O format, 010 or 001 indicate a memory-to-register movement instruction. The format shown in the example above is a move data format that moves data between memory and registers. In this format the FORMAT SELECT and OPERATION CODE must always be 001 or 010. When this value is represented in the first 3 bits of an instruction word, the other bits are interpreted as the example shows.

2.  AR specifies one of two arithmetic registers.

3.  IND specifies indirect addressing when the bit is set to 1.

4.  IR specifies an index register to be used in address formation.

5.  ADDRESS represents a location in memory. This value plus the value in an index register will form a complete memory address. Notice the short address field in the instruction. This suggests that the address is usually added to other values to be able to address more than 256 bits of memory.

## 5.3.  Other Formats

Both of the above examples are from machines that have one size of instruction. To enrich the operation set, one of these machines, using a short instruction word, has different formats to represent different classes of operation.

As we saw in the discussion of the VAX-11, there are architectures that use different size instructions for different functions. The size of an instruction usually depends on the addressing conventions an instruction uses. Thus an instruction that references two memory locations will be longer than an instruction that references one register and one memory location or two registers as the source or target of an operation. Like the VAX-11, the INTEL 8080 and IBM Series/1 have an architecture that uses variable instruction sizes depending on addressing requirements.

Thus in the INTEL 8080 the basic instruction is a single 8-bit word that carries the operation code or the operation code and one or two register addresses. The referenced registers contain a 16-bit memory address of which the 8 high-order bits are in a register H and the 8 low-order bits are in a register L. References to memory can be made in the one-word format by setting address values in these specific registers prior to the execution of the instruction.

INTEL 8080 immediate instructions use a two-word format in which the second word contains 8 bits of immediate data. The set of instructions that include addresses to memory uses a three-word format of 24 bits of which two words contain a direct memory address.

The System/370 generic architecture also uses instructions of various sizes, depending on addressing. In this architecture, however, any particular instruction has a fixed size depending on whether it is an R–R, R–M, or M–M instruction. The addressing mode is coded in the operation code. This contrasts with the VAX-11 in which only the length or type of data is coded into the operation. Each operand address type is uniquely described and is whatever length is necessary; there is a variable number of addresses.

In all machines in which the instruction size may vary, the size of the instruction must be made known in some way to the control circuits that advance through the sequence of a program. Instead of adding a 1 to an instruction counter to develop the address of the next instruction, it is necessary to add the length of the current instruction. This length may be coded in the operation code representation of the instruction. If a particular instruction operation code may vary in length, then some sort of byte used counter must be developed to form the address of the next instruction in memory.

Just as it is possible to use instructions of variable size involving one or more words to form an instruction, it is possible to represent more than one instruction in a computer word. The CDC Cyber architecture and the architecture of the Cray Research CRAY-1 have variable size instructions, multiples of which may fit into a single word. The CRAY-1 instruction may be either 16 or 32 bits long. The 16-bit instruction format for arithmetic and logical instructions includes a 7-bit operation code and three 3-bit register designators indicating the source of two operands in a register and a register in which to place the result of the operation. When immediate data or references to memory are made, a 32-bit format is used. This format has a 4-bit operation code, a 3-bit index register address, a 3-bit arithmetic register indicator, and a 22-bit memory address. A 64-bit word can contain four 16-bit instructions or two 32-bit instructions. In the CRAY-1 instruction, parts may span across word address boundaries.

## QUESTIONS

1. What are the truth tables for OR, AND, NOR, and NAND?

2. What is the purpose of branching instructions?

3. What is a privileged instruction?

4. What is the concept of a symmetric instruction set?

5. How does UNIVAC 1100 provide for field selection within a 36-bit word?

6. How does NOVA use the concept of format selection?

7. Describe reasons for variations in the size of VAX-11 instructions.

# Chapter 7

# Changes in
# Program Sequencing

## 1. SEQUENCE ALTERATION

A program contains a list of instructions in sequential memory locations that are executed sequentially by the augmentation of a control counter that points to a successor instruction. There are three basic ways of altering instruction sequence: branching, discussed in this chapter; subroutine linkage, discussed in Chapter 8; and interrupts and exceptions, discussed in Chapter 9.

## 2. BRANCHING

All computer architectures contain a set of instructions that will change the sequence of instruction execution either *unconditionally* or *conditionally*. Unconditionally means that the next instruction to be executed will be taken from the address of the branching instruction. Conditionally means that the address associated with the branch instruction will contain the next instruction if a certain condition, stated by the branch instruction, is found to be true. Otherwise the instruction that follows the branch instruction will be executed.

A partial listing of branch conditions may be as follows:

1. Operand equal to operand: IF A=B THEN
2. Operand unequal to operand: IF A≠B THEN
3. Operand greater than operand: IF A>B THEN
4. Operand smaller than operand: IF A<B THEN
5. Operand is negative: IF A<0 THEN
6. Operand is 0: IF A=0 THEN
7. Operand value lies between range: IF A>B & A<C

The conditions for branch are explicitly stated or implied by IF statements in the programming language. The actual machine coding will differ in ways that are not apparent in the programming language due to:

1.  The specifics of the relationships between operation codes and addressing forms. These will determine what addressing forms a branching operation may use, whether comparands must be in registers or in memory, and whether the forms of the branching operation change as a result of where comparands are.

2.  The data representation of comparands. Machines with multiple data representations may have distinct branching operations for each data representation.

3.  Whether both the concept of relative magnitude of two comparands and the concept of a single value being greater than, equal to, or less than 0 is represented in the operation codes.

Two fundamental concepts are represented in the table of IF conditions given above. Some of the conditions concern the relative magnitude of two comparands, some of the conditions specify a particular status of a given data element. Architectures commonly, but not universally, distinguish between the two forms.

The branch function involves a statement of the condition that will cause a branch and up to three addresses. The three addresses indicate comparands and a branch address if the condition is true. The *fall through* address is universally implied by the contents of the instruction counter. In some cases only one comparand is required because a comparison is made on the status of a single value. That is, instead of comparing a value to a constant of 0 there is a direct test on a single operand for a 0 value. The details of branch logic architecture reflect different decisions about how to effectively set conditions or specify operands within the context of the addressing conventions of the architecture.

One form of branch logic may involve a three-address branch that defines the condition, names two comparands, and gives a branch address. Many architectures, however, use a sequence of instructions to set up a branch point. The exact sequence depends on whether a relative magnitude comparison is being made between two comparands or whether a condition of a particular data value is being tested. In many architectures the sequence will also depend on previous computation and on whether the values to be tested are already in registers. The addressing conventions that indicate whether branching type instructions can refer to memory for all or some operands, or must have comparands in registers, influence the details of a branch sequence.

The data representation of an architecture will also affect the branching logic since it is necessary to indicate whether pure binary or binary-coded decimal representation is used for comparands. It is also possible in some architectures to indicate whether comparison for relative magnitude is to be algebraic or based on absolute (non-signed) values.

LOAD R1, Address 1
BRANCH ON ZERO R1, Address ?

The above places a value in an operand register and inspects the value for 0. The implication of the example is that 0 tests cannot be performed in memory, and a value must be brought to a register before a 0 test can be applied. This may be true because a memory address cannot be specified in a BRANCH ON ZERO instruction, that is, a BRANCH ON ZERO must be in the R-M form.

LOAD R1, Address 1
SUB    R1, Address 2
BRANCH NEGATIVE R1, Address 3

The above determines whether register 1 has become negative after a subtraction. This might be efficient if the subtract operation is in any case required for program flow. It is an inefficient form if the only function of the subtraction is to determine if the value in address 2 is greater than the value in address 1.

An alternate form for comparing relative magnitude would be as follows:

LOAD R1, Address 1
COMPARE-DECIMAL R1, Address 2
BRANCH LOW, R1, Address 3

The relative efficiency of this depends on whether the COMPARE instruction is faster on a particular machine organization than an arithmetic instruction. Of course, as we have seen, the relative efficiency of any R-M instruction sequence depends on whether the preliminary loads of a register are needed or whether the value exists in the register as a result of previous computation.

A fundamental variation in branch logic architecture derives from the use of a condition code control register. Consider:

LOAD R1, Address 1
LOAD R2, Address 2
JUMP HIGH, R1, R2, Address 3

LOAD R1, Address 1
LOAD R2, Address 2
COMP-BINARY, R1, R2
BRANCH HIGH, Address 3

In the first sequence the JUMP instruction performs the test for relative magnitude in the two registers and then does or does not cause a branching. In the second sequence the COMPARE instruction sets a code for the BRANCH instruction to test. The relative efficiency of these two forms depends on general aspects of the addressing flexibility of the architecture and the sequence of events that precedes the branch point.

## 3.  CONDITION CODE LOGIC

As suggested above, an approach to conditional branches is called *condition code logic*. Somewhere in the architecture, either as a separate control register or as part of a larger control structure, there is a set of bits that represent conditions of data. The S/370 generic architecture has a control register, discussed in detail in Chapter 9, called the Program Status Word (PSW) that performs various control functions. Two bits of this word are used to hold condition codes. These 2 bits may assume four unique values. The setting of these values may be accomplished by a large number of instructions in the S/370 architecture. All of the arithmetic instructions, the logical (boolean) instructions, many I/O instructions, most register load instructions, most shift instructions, and some control (privileged) instructions may result in a new setting for the condition code in the PSW.

For example, an addition may result in various conditions of data after it is performed. The result may be 0, less than 0, greater than 0, or an overflow. After an addition the condition code in the PSW will read:

| | |
|---|---|
| 00 | Value is 0. |
| 01 | Value is greater than 0. |
| 10 | Value is less than 0. |
| 11 | Overflow has occurred. |

Similarly, an instruction executing a boolean AND will leave the condition code with a value of 00 or 01 depending on whether or not the resulting operation left any 1 bits in the result.

The function of the Branch on Condition (BC) instructions is to test the value of the condition code and to cause or not cause a branch, depending on whether a particular value is found in the condition codes. The condition being tested is described in a field of the Branch on Condition instruction called the *mask*. The mask determines whether the branch will be taken on condition code values of 0, 1, 2, or 3. For example, if the mask of the instruction reads:

Bit Designation in Mask

| 0 | 1 | 2 | 3 | |
|---|---|---|---|---|
| 1 | 0 | 0 | 0 | A branch will occur if condition code is 0 |
| 0 | 1 | 0 | 0 | A branch will occur if condition code is 1 |
| 0 | 0 | 1 | 0 | A branch will occur if condition code is 2 |
| 0 | 0 | 0 | 1 | A branch will occur if condition code is 3 |

The specific process (on 370 architecture) is as follows: A condition code value will select a particular bit position in the mask. If this bit position is 0 the branch does not occur. If it is 1 the branch does occur. Thus, condition code 00 selects bit 0 in the mask. Bit 0 in the mask means that a branch will be taken if the condition code indicates a 0 value. If the branch instruction is not interested in 0 values then bit 0 in the mask will not be on and the 0 value indicator in the condition code will not affect program sequence. If the bit is on but the condition code is not 00 no branch will occur because the condition indicated by the mask does not obtain. Condition code 01 (value less than 0) selects bit 1 of the mask; condition code 10 selects bit 2; condition code 11 selects bit 3. More complex condition definitions can be obtained by using more than 1 bit in the mask. Thus, combinations like 0 or greater can be defined. All of the above examples relate to testing the condition of a value developed in a register as a result of a load or an operation on the value.

| Condition Occurs | Condition Code | Mask | Branch |
|---|---|---|---|
| Value in Register = 0 | 00 | 1000 | YES |
| Value in Register = 0 | 00 | 0010 | NO |
| Value in Register > 0 | 01 | 0100 | YES |
| Value in Register > 0 | 01 | 1000 | NO |
| Value in Register < 0 | 10 | 0010 | YES |
| Value in Register < 0 | 10 | 1100 | NO |

In earlier chapters we mentioned that assembly language designers have some degree of freedom in determining the representation of the instruction set of the architecture. The assembly language for the S/370 architecture presents the instruction set as having a number of branch on condition instructions. These include a set of branches to be used after arithmetic and load instructions and are written in assembly language as:

| Mnemonic | Condition |
|---|---|
| BH | Branch on overflow |
| BP | Branch on plus |
| BM | Branch on minus |
| BNP | Branch on not plus |
| BNM | Branch on not minus |
| BNZ | Branch on not zero |
| BZ | Branch on zero |
| BNO | Branch on no overflow |

Each of these instructions has the same operation code. The differences between them lie in the coding of the condition definition mask. The branch addresses may be taken from a register or a memory address formed by using the base register, index register, and displacement field addition used to form all addresses. That is, branching may use R–R or R–M formats.

Any load or computational instruction will set the condition code. The intent of setting condition codes with so many instructions is to allow branching to take place as a direct result of the operation of a previous instruction. Since an arithmetic instruction might in any case be performed on a value, branching on a condition established by the arithmetic is very efficient. Of course, coding that does the arithmetic only to establish the branch condition would be less efficient.

An alternative way of setting condition codes is by using a COMPARE instruction. This instruction will compare two values and determine whether they are equal or whether the first operand is low or high. It sets condition codes 0, 1, and 2 in accordance. The Branch on Condition instructions then test the condition code. The COMPARE instruction may compare values in registers, a value in a register with a value in memory, or two values in memory. In addition, there are variations in the comparison that describe the length of the operands (halfword, word, or doubleword) and whether the comparison should be arithmetic or logical, and there are comparisons for various forms of data representation.

The same Branch on Condition instructions that were used above for determining arithmetic conditions are used to test the results of a comparison. In the assembler, however, a unique set of mnemonics is provided to relieve the programmer from concern about establishing mask values. These instructions are as follows:

| | |
|-----|-------------------|
| BH  | Branch high       |
| BL  | Branch low        |
| BNH | Branch not high   |
| BNL | Branch not low    |
| BE  | Branch equal      |

The VAX-11 also represents condition codes in a PSW. The details of the VAX-11 PSW differ from that of the 370. In this architecture there are 4 condition code bits, each with a special function. Thus, there is a *negative bit* that indicates that the result of a previous instruction is less than zero. There is a *zero bit* that indicates the result of a previous instruction is zero. There is an *overflow bit* indicating that the result of the previous instruction caused an overflow by trying to represent a value too large for the data type being used. There is a *carry bit* that indicates a carry out of the most significant bit of data.

As with the S/370, a number of instructions other than the comparison instructions may affect the value of the condition codes. The architecture contains a set of branch instructions that are classified as *signed* and *unsigned* branch instructions. The signed instructions test for equality and inequality between signed operands (arithmetic magnitude is being tested), the unsigned instructions test for equality and inequality between unsigned integers (absolute magnitude is being tested). Comparands may be 1, 2, or 4-byte fields in floating point, binary, or coded binary form.

Addressing for branches on the VAX-11 is limited to a one-byte displacement that is added to the instruction counter. Branches are therefore all relative, and the transfer address is stated as being some distance away from the current value of the location counter, plus or minus. For branch addresses farther than that distance from the instruction counter an instruction exists that will allow branching using a 16-bit signed displacement. An alternate branch instruction called a JUMP provides for transfers using any of the addressing forms available on the VAX-11.

Many other architectures use some form of condition code setting to support conditional changes in program flow.

## 4. SKIP LOGIC

The challenge in conditional logic definition is how to handle the problem of three necessary addresses. Condition code architecture uses two instructions. There is some efficiency in this because condition codes may be set by arithmetic operations so that a BRANCH need not always be preceded by a COMPARE. Another approach, called *skip logic*, is to imply the conditional branch address. Thus a SKIP instruction tests for a particular condition, and, if it obtains, the next instruction will be taken from the location immediately following the SKIP. If it does not obtain, the next instruction will be taken from the location two locations beyond the SKIP. The location just after the SKIP would characteristically contain an unconditional branch.

SKIPs may be associated with condition code settings, but there are a number of architectures where SKIP logic is employed that do not use condition codes. The status of the processor or other system element being tested is not made available in an organized condition code register of the architecture. Thus, testing of the condition code is not accomplished by a separate instruction but is inherent in the determination of the presence of the SKIP condition.

> LOAD R1, Address 1
> SUB R1, Address 2
> SKIP-ON-NEGATIVE R1
> JUMP Address 3
> Fall Through

In the above the SKIP instruction tests for a negative value in register 1. If register 1 is not negative the next instruction to be executed is an unconditional jump; if register 1 is negative the instruction at the fall through location is executed.

An example of a popular SKIP logic machine is the Data General NOVA architecture. In this architecture skipping is implemented in two forms. There are SKIP instructions that test for particular conditions defined for each SKIP instruction. For example, there is a SKIP that tests for the busy status of an I/O device. It is possible, however, to specify a condition for skipping within the context of performing another instruction. This kind of skipping is referred to as the SKIP function within an arithmetic or logical instruction.

The arithmetic instructions of the NOVA architecture can specify a SKIP function as a suboperation. The assembly language coding looks like:

ADD 1,2, SZR

This instruction specifies that the contents of register 1 are to be added to the contents of register 2 with the results placed in register 2. If the result of the arithmetic operation is 0 then the next instruction should be skipped. The usual contents of the next instruction, executed if the SKIP condition is not found to exist, is an unconditional JUMP instruction.

The SKIP conditions that can be specified as subfunctions of another instruction are:

Never skip.
Always skip.
Skip on 0 carry.
Skip on non-0 carry.
Skip on 0 result.
Skip on non-0 result.
Skip if either carry or result is 0.
Skip if both carry and result are non-0.

An architectural objection to the SKIP instruction derives from the necessity for always providing an unconditional JUMP following the SKIP instruction. This is an inefficient way of providing a branch address since it requires space for an operation code. However, when skipping can be done as a subfunction of arithmetic operations that are to be performed anyway, the unconditional JUMP does not really require more space than a BRANCH testing a condition code.

Most architectures pay particular attention to branching as the result of some value in an index register, frequently combining index register modification with a test for a 0 condition. Details vary on how the index register is modified, whether the modification is done before or after the test, and the complexity of the condition being tested, but it is just about universal to have a set of specialized index register modification and conditional branching instructions to facilitate loop control and address advance.

A processor spends between 25 and 35% of its time handling conditional or unconditional branches. Efficient architecture and organization in this area are very important. One must be aware of all of the associated required instructions, in addition to the actual transfer instruction, to determine the efficiency of any architecture.

## QUESTIONS

1. What is the purpose of a conditional branch?

2. What is a condition code?

3. What is the possible efficiency of setting condition codes by instructions other than COMPARE?

4. What relationship does condition code setting have with the concept of a regular architecture?

5. What is the relationship between PSW and condition codes in some architectures?

6. What is SKIP logic?

7. Is it necessary to have SKIP instructions in a SKIP architecture?

# Chapter 8

# Subroutine Linkage

## 1. CONCEPT OF A SUBROUTINE

In the construction of a program a programmer often finds that there is a computation that must be made at various times in the flow of program logic. Rather than repeat the statements that represent this computation at each point where it is necessary, the programmer may form the computation into a single set of statements, provide the set of statements with a name of its own, and have the program flow refer to the single program whenever that particular computation is required.

The set of statements representing the frequently used program is commonly called a *subroutine* or a *procedure*. The language conventions that determine how the main stream of statements will refer to the subroutine are called *subroutine linkage* or *procedure invocation* conventions. These conventions determine the manner in which the subprogram will be invoked, how values will be passed to it, how it will return to the invoking program, and how it will pass the results of its computation back to the main stream.

Languages differ in the details of subroutine and procedure invocation, and the variations in concept between FORTRAN and PASCAL, for example, can be important. There are also some differences in concept between subroutines and procedures in various languages and between various languages.

Block structure languages, in general, define a hierarchical structure, built around notions of nested BEGIN and END statements, that provides a relationship between "inner" and "outer" procedures as regards the sharing of variable names (global variables). Implementation of procedure invocations must observe scope of name rules defined by the hierarchical relationships between inner and outer procedures. The subroutine call may not imply all of the memory allocation and linkage issues that a procedure call implies. Students of languages and language processors must be very familiar with variations in the details of program and subroutine, procedure, and function relationships.

What is common to all linkage is the necessity for naming the program to be invoked, passing values (parameters) that serve as the input to the called program, and providing for the output of the called program to be returned to the caller. In this chapter we limit our discussion to the fundamental architectural features of hardware that may be used to efficiently support the notion of a program calling another program and having the called program return to the calling program, providing a value or set of values upon return. The architecture must address the issues of how the invocation and return are made, how parameters are passed to the subprogram, and how variables used by the calling program may be accessed by the called program.

Throughout most of the chapter we deal with the abstraction of a subroutine and do not distinguish between subroutine calls and procedure invocations. The last section makes some comments about issues of addressing involved in procedure calls. This simplification is reasonable in an introductory hardware book, but the reader should be aware that the topics of linkage and procedure invocation are highly specialized and are only introduced in this chapter. Issues of addressing conventions, protection, segmentation, and operating systems design are all relevant.

## 2. SUBROUTINE LINKAGE

Let us assume that, whether or not the verbs CALL and RETURN are explicitly used in the high-level language, at some point in compilation an intermediate form CALL will be generated when a subprogram is invoked. Also the subprogram will have a RETURN macro generated in intermediate form. Therefore, the hardware architectural issue is what should be generated as code when CALL and RETURN are recognized by the code generator routines of the compiler.

The invocation of a subroutine involves conventions about parameter passing, result or condition return, saving and restoration of registers, and addressing ranges of calling and called programs that may or may not be directly addressed by the architecture.

Linkage can be accomplished by instructions and by addressing conventions built around sequences of general-purpose instructions, such as LOAD and STORE, or they can be supported by instructions and structures that make CALL and RETURN efficient. Efficiency means that the number of instructions executed to accomplish a linkage is minimized.

## 3. BASIC LINKAGE INSTRUCTION

A fundamental and ubiquitous feature of an architecture is the *BRANCH-LINK* or *BRANCH-RETURN* type instruction to support subroutine calls. Almost all machines, beginning with the first UNIVAC, have a concept of executing a branch and recording, as part of the function of the branch, the contents of the instruction counter at the time of the execution of the branch. The contents of the instruction counter indicate the next sequential instruction, the instruction to be executed after the subroutine has completed. This address is called the *return point*.

Variations in the BRANCH-LINK concept involve specific conventions for where the return address is recorded. A subroutine call has the general form:

| Symbolic Location | Instruction/Comment |
|---|---|
| CALLPOINT | BRANCH-LINK subroutine A |
| CALLPOINT+1 | Resume calling program |

When the BRANCH-LINK is executed from location CALLPOINT, the address CALLPOINT+1 will be recorded in the system. Address CALL-POINT+1 is the address in the instruction counter representing the beginning of the instruction that follows the BRANCH-LINK. (In byte addressable machines the actual value of CALLPOINT+1 will be the position of the start of the BRANCH-LINK plus the number of bytes the BRANCH-LINK occupies.) When subroutine A is complete it will return to the caller by executing an instruction that will cause the return point to be placed in the instruction counter. This will cause the instruction at CALLPOINT+1 to be executed.

The return address (CALLPOINT+1) may be recorded in various places:

1. A fixed address in memory that is always used to hold the instruction counter when a BRANCH-LINK is executed. This may be a specialized location at the top or bottom of the memory addresses.
2. An address in memory specified by the BRANCH-LINK instruction.
3. A register specified by the BRANCH-LINK instruction. This may be an operand register or one of a set of addressable control registers used for linkage.
4. A specialized register implied by the BRANCH-LINK instruction.

Systems conventions associated with compilation and compiler libraries are always established so that called subroutines know where to find the return point.

One ancient convention is to put the return address into the first location in the called subroutine. The compiler knows when it is compiling a subroutine and provides for the first location to hold a return address that will be placed in it as a result of a BRANCH-LINK. The instructions of the subroutine start at the second location. A BRANCH-LINK to subroutine A will then record CALLPOINT+1 in the location subroutine A and really branch to subroutine A+1. When ready to return, subroutine A knows that the return point is in its own first location. If the architecture has indirect addressing, a BRANCH to the first location using indirect addressing will accomplish the return. If no indirect addressing is available, the first location may be placed in a register and a BRANCH REGISTER will accomplish the return. Some architectures provide for the hardware fabrication of a BRANCH instruction to be placed in the location holding the return address. The subroutine then executes two simple branches, a BRANCH to the first location followed by a BRANCH to the return point.

A simple example, where indirect addressing is used:

| Location | Symbolic Loc | Contents/Comment |
|----------|--------------|------------------|
| 050 | CALLPOINT | BRANCH-LINK SUBA (BRLNK 1000) |
| 051 | CALLPOINT+1 | Continue calling program |
| . | | |
| . | | |
| . | | |
| 1000 | SUBA | 051/Recorded by execution of BRANCH-LINK |
| 1001 | SUBA+1 | First instruction of subroutine A |
| . | | |
| . | | |
| . | | |
| 1456 | ENDA | BRANCH* SUBA(BR * 1000)Indirect BRANCH to return point in SUBA |

Another convention involves the use of a specified register where the return address is placed and the return to a caller is accomplished by the execution of a BRANCH that indicates that the branch address is in the register.

| Location | Symbolic Loc | Contents/Comment |
|----------|--------------|------------------|
| 050 | CALLPOINT | BRANCH-LINK, REG1, SUBA |
| 051 | CALLPOINT+1 | Next instruction of calling program |
| . | | |
| . | | |
| . | | |
| 1000 | SUBA | First instruction of subroutine A |
| . | | |
| . | | |
| 1450 | ENDA | BRANCH REGISTER, REG1/ Branch to address in register 1. |

One of the possible requirements if a register is used to pass the return address is that the called subroutine will have to store the return address to free the register and then restore it to the register before the return BRANCH is issued. The called program, of course, has the option of placing the return address in any convenient memory location and referring to it in the most convenient way.

The BRANCH-LINK may also use a specified memory location in which to record the return point. To do this it is necessary for there to be a defined set of addresses that are accessible to both the calling and the called programs.

## 4. BASIC STACK LINKAGE CONTROL

A form that is rather popular in recent architectures is the use of a stack as a mechanism for holding return addresses. On execution of a BRANCH-LINK instruction, the instruction counter is placed in a stack in memory that is indicated by a top of stack pointer. A return from a subroutine is accomplished by issuing an instruction that pops the top of stack to the instruction counter. The concept is very similar to the concept of operand stacks we discussed previously except that the stack is defined in memory instead of in the registers.

Our discussion of operand stacks involved a LIFO relationship defined between the arithmetic registers and a stack pointer. In the discussion of operand stacks, the use of memory locations for a stack was necessary only when the stack became too long to be represented in the register population.

It is usual that the stacks defined to support subroutine linkage are defined in the general memory and the stack is defined by a pointer containing a memory address that indicates the top of the stack. The definition of stack areas may be represented in the architecture by stack segment registers. Manipulations of the stack then use addresses that are offsets of the stack area defined by a stack segment register.

The most fundamental use of a stack mechanism to support subroutine linkage is as a structure to represent return addresses. The stack provides a place to put a return address rather than in a general register. Thus, in an architecture that used a basic stack linkage mechanism, the compiler would pass parameters outside the context of the stack (by loading registers, etc.), but the BRANCH-LINK instruction would be replaced by a CALL instruction that placed the return point on the top of the *linkage stack*. This linkage stack is a LIFO list of return points that is pushed when a CALL is executed. When the subroutine completed it would return by issuing a RETURN, which would use the address at the top of the stack as a branch address.

The Intel 8080 provides an example of a basic subroutine call linkage stack mechanism. The architecture contains several CALL instructions that allow for unconditional or conditional subroutine calls. The address of the called subroutine is provided in the CALL instruction. When the CALL instruction is executed the contents of the instruction counter (which is the address of the instruction following the CALL instruction) is placed onto the stack. The subroutine issues a RETURN instruction to reenter program flow. The RETURN instruction places the return address placed on the stack by the CALL instruction back into the location counter. In effect, the CALL instruction is a stack PUSH operation and the RETURN is a stack POP operation. The special property of CALL and RETURN is that the value placed on the stack or taken from the stack is coming from or going to the instruction counter.

The memory area used for the stack must be indicated by a stack address pointer. The 8080 architecture contains a special 16-bit register called a *stack pointer register*. The pointer placed in this register implies two locations for every CALL and RETURN operation. The most significant 8 bits of the instruction counter are stored at the address one less than the address indicated by the stack pointer. The least significant 8 bits are stored in the address two less than the address indicated by the stack pointer. Thus, a contiguous 16-bit entry in memory is defined by the stack pointer. As part of each CALL operation the value in the stack pointer register is automatically decreased by two. On a RETURN instruction the value in the stack pointer register is automatically increased by two, and a 16-bit value is transferred from the stack to the instruction counter.

The table below shows the behavior of the CALL and RETURN instructions. We assume a called subroutine is at location 5000. The CALL instruction is a 3-byte instruction, the RETURN instruction is a 1-byte instruction. The instruction after the CALL instruction is at location 4002. The stack pointer register has previously been loaded. During the execution of the subroutine at 5000 there is an additional subroutine call to a subroutine at 6000. This is called a *nested* subroutine call. The stack mechanism provides for the nesting of subroutine calls of indefinite depth.

| IC | Instruction | Stack Pointer | Stack at End of Instruction |
|----|-------------|---------------|-----------------------------|
| 4000 | CALL 5000 | 1000 | 998 last 8 bits of value 4002 |
|  |  |  | 999 first 8 bits of value 4002 |
| 4002 | RETURN POINT |  | N/A |
| 5000 | Subroutine 1 | 998 | 998 last 8 bits of value 4002 |
|  |  |  | 999 first 8 bits of value 4002 |
| 5030 | CALL 6000 | 996 | 996 last 8 bits of value 5032 |
|  |  |  | 997 first 8 bits of value 5032 |
|  |  |  | 998 last 8 bits of value 4002 |
|  |  |  | 999 first 8 bits of value 4002 |
| 5032 | RETURN TO subroutine 1 | N/A | |
| 6000 | Subroutine 2 | 996 | 996 last 8 bits of value 5032 |
|  |  |  | 997 first 8 bits of value 5032 |
|  |  |  | 998 last 8 bits of value 4002 |
|  |  |  | 999 first 8 bits of value 4002 |
| 6100 | RETURN | 998 | 998 last 8 bits of value 4002 |
|  |  |  | 999 last 8 bits of value 4002 |
| 5032 | Subroutine 1 | 998 | 998 last 8 bits of value 4002 |
|  |  |  | 999 first 8 bits of value 4002 |
| 5076 | RETURN | 1000 | Stack empty |
| 4002 | RESUME | 1000 | Stack empty |

The stack always contains the return address of the calling routine.

In the above, the first subroutine call is issued at location 4000. Since the CALL is a 3-byte instruction the next instruction will begin at location 4002. 4002 is therefore the return point for this subroutine call. In locations 999 and 998 the 16-bit value 4002 is formed as a single stack entry. When SUBROUTINE 1 is running it calls SUBROUTINE 2. This call is made by a 3-byte instruction beginning at 5030. The return address is 5032. This entry is placed in locations 996 and 997 as a stack entry. When SUBROUTINE 2 returns to SUBROUTINE 1, the value 5032 is popped from the stack. The top entry in the stack is again 4002, representing the return point to the initial caller.

Many other architectures are similar to this. Some contain a limit on the depth of a linkage stack, and there are variations in how stack pointers are modified and where they are held.

## 5.  PARAMETER PASSING

The above sections did not consider much of the complexity of subroutine call and return since they did not include discussion of the means by which a calling program sends parameters to a called program, or a called program returns values to a calling program.

Consider the simple CALL statement, CALL PROGRAM A (X, Y, Z). It will be encountered by the compiler while compiling a Program B that is independent of the compilation of A. The compiler may generate instructions of the following type for a machine that has no formal architecture for the support of subroutine calls beyond a basic linkage stack.

1.  LOAD R1 X
2.  LOAD R2 Y
3.  LOAD R3 Z
4.  CALL *PROGA
5.  Etc.

The above coding may place actual values or the addresses of values in the registers. Various compilers and systems environments use different conventions about how parameters are passed and whether values themselves are placed in registers or the addresses of values held in memory. Conventions about the use of registers, their saving, and their restoration naturally apply whenever registers are used as a program-to-program communication mechanism.

The compilation of Program A will encounter some statement indicating that the program expects parameters on being called. This statement will have a form similar to PROGRAM A: (X,Y,Z). X,Y,Z will naturally be declared according to form and mode in a declaration in each program. If this call provides values, a reasonable compilation for A would be as follows:

1. STORE R1, P1
2. STORE R2, P2
3. STORE R3, P3
4. Etc.
5. Etc.
6. LOAD R4 RESULT
7. RETURN

The first instructions store the parameters in locations in the called subroutine A. Then A operates until it develops the value desired by the caller B. It places the value in a register and issues a RETURN that pops the return point from the stack to the instruction counter.

In several architectures the multiple LOAD and STORE instructions may be replaced by a single LOAD MULTIPLE and STORE MULTIPLE instruction that establishes values or takes values from a set of registers. The compiler must organize parameters in memory so that values may be taken from contiguous memory locations or placed in contiguous locations when the multiple register instructions are used.

## 6. ACTIVATION RECORD

A subroutine call, besides requiring a convention for the passing of parameters, may also require the provision of a working area the called routine may use when it is invoked.

The declared local variables of a subroutine must be independently allocated space every time the subroutine is called. This is necessary because of possible *recursion* and *reentrance*. Recursion is the phenomenon of a program calling itself ("to iterate is human, to recurse divine"). Reentrance provides for the possibility that in a shared system several callers may share the same subroutine and issue calls to it. The subroutine must be fresh each time it is called and generate values unique to each call. In the environments developed by many operating systems, it is possible for a subroutine to be suspended as a result of a task switch before it is completed. At the time of the suspension a set of values associated with the subroutine must be maintained so that the subroutine may be later resumed. It is possible that the subroutine may be called by the task which has received control of the processor. A completely unique and separate working area must be established for the second call of the subroutine. The code of the subroutine must have some way of properly addressing the working area that is appropriate.

An *activation record* is a space allocated to a subroutine each time it is called so that values unique to the call will have a unique place in memory associated with a particular call of the subroutine.

## 7. STACKS, PARAMETERS, AND ACTIVATION RECORDS

A modest generalization of the stack concept enables a stacking mechanism to be used to support the requirements of parameter passing and activation record definition. The single stack entry for the return address is generalized so that an area of variable size may be acquired by a subroutine for use as a working area, a place to store parameters, and a place for the return address.

An example of the use of the stack concept in enhanced support for subroutine linkage can be found in both the VAX-11 architecture and the Series/1 architecture. We briefly describe the nature of the Series/1 stack mechanism in connection with subroutine linkage support.

The definition of an area in memory as a stack is accomplished by the formation of a *stack control block*. A stack control block is an element containing three Series/1 16-bit words. One word contains the address of the lowest memory location to be part of the stack. Another contains the address that is the highest address to be part of the stack. A third word contains the address of the current top of the stack. Together the three words define the length of a contiguous area in memory to be used as a stack and the current top of the stack. An empty stack has the same address for the high limit address and for the top of stack address. Stacks grow from high memory locations to lower memory locations (toward address 0).

### Stack Control Block

| | |
|---|---|
| Word 1 | Lower bound of stack |
| Word 2 | Higher bound of stack |
| Word 3 | Current top of stack |

A subroutine call is made by using a BRANCH-LINK type of instruction that addresses the subroutine and places the return address in a general-purpose register, the specific register is established by systems convention and not built into hardware. Also by convention, the address of the stack control block to be used to address the stack is provided in a register and the address of a parameter list in another.

### Register Contents (Register Designation Symbolic)

| | |
|---|---|
| Register return | Return address |
| Register stack | Address of stack control block |
| Register parameter | Address of parameter list |

The first instruction executed by the called subroutine is a STORE MULTIPLE instruction. This instruction specifies the registers that are to be stored on the stack. The Series/1 convention is that the called subroutine is responsible for storing the registers it must use during its own execution and restoring them to the values that were in them at the time of the call when the subroutine is completed. The registers that MUST be stored are Register Return, Register Stack, and Register Parameter.

The execution of the STORE MULTIPLE instruction places the values of this set of registers, and any others indicated in the STORE MULTIPLE instruction, on the stack. The instruction uses Register stack containing the address of the stack control block to gain addressability to the stack. An entry in the stack control block, remember, provides current top of stack.

Besides operating as a kind of PUSH to place registers on the stack, the STORE MULTIPLE instruction also specifies the size of an area that is to be allocated on the stack as a working area for this call of the subroutine. At the end of the STORE MULTIPLE instruction, the top of stack pointer in the stack control block is lowered to reflect the number of registers stored and the size of the working area allocated on the stack for this activation. A pointer to the working area is placed in the highest addressed register stored by the subroutine.

The table below shows the status of a Series/1 stack entry as a result of a subroutine call to a routine that requires 20 words of working area while it operates and that saves two registers besides the parameter list and return address registers. The base of the stack is at address 5000, no stack entry may go beyond 3000, and the entire activation record for this call contains 24 locations. The new top of the stack is at location 4976. At the beginning of the STORE MULTIPLE instruction, the top of stack address was 5000.

| Stack Control Block | Stack |
|---|---|
| TOP OF STACK: 04976 | 04976: Control word |
| HIGH ADDRESS: 05000 | 04977: First working area word |
| LOW ADDRESS: 03000 | 04996: Last working area word |
| | 04997: Return address |
| | 04998: Parameter list address |
| | 04999: Saved register |
| | 05000: Saved register |

When the subroutine is complete a LOAD MULTIPLE and BRANCH instruction is executed. This is effectively a RETURN instruction. It causes the reloading of saved registers and the downward adjustment of the top of stack pointer in the stack control block. The working area used by the subroutine is effectively destroyed.

This architecture reflects the spirit of several architectures that use the stack concept for activation record and parameter passing as well as for linkage. Some architectures provide for the working area to be noncontiguous with the stack area and a LINK mechanism that points to the working area from a position in the stack. Where these features do not exist it is still possible for a subroutine to acquire additional noncontiguous storage for a particular activation.

Any interruption of the subroutine will leave the activation record defined in the stack intact. A new call will create a stack entry "above" the activation record for the previous call. When the second call completes, its activation record will be removed from the stack and the activation record of the first call will be at the top of the stack.

Some operating system and compiler combinations use the concept of an activation record in other ways. A calling program, for example, may put parameters for a called program in its own activation record and provide an address to the called program. The called program then has a controlled addressability to the activation record of its caller as well as to its own activation record.

Stacking mechanisms often have certain instructions to manipulate the stack on the occurrence of certain abnormal events. An example of such an event would be an error in a called program such that it returned to its caller with an error indication. The activation record would be left on the stack so that the caller or some other routine might undertake some problem determination activity. At the end of the error analysis the activation record would be explicitly popped from the stack.

It is important that the reader distinguish an activation record from an operating system control block. The activation record contains parameters, linkage addresses, and variables. The program control block is used for register values, resume addresses after an interrupt, and, usually, to hold pointers to various resources that the program may address (e.g., files and devices). There is clearly some intersection between the two objects, but so far they have usually remained unique in systems structures. The activation records are used for programs within a process. The program control block is used to control the status of processes.

## 8. ADDRESSING AND LINKAGE

An aspect of subroutine linkage or procedure invocation lies within the context of addressing and addressing conventions. A particular issue is the addressing context within which a procedure or a subroutine will run and the extent to which variables available to a calling program will be available to a called program.

In some architectures the subroutine may exist in memory in such a way that it is addressed using the same base or segment register that is used by the calling program. It is able to address global variables and parameters in a straightforward way because it can address all the addresses that the caller can. A problem exists if it is desired to limit the addressability of the called program to those variables declared global variables by the caller. This is very difficult to do. Protection keys may not give enough granularity to provide addressing only to global variables. The problem must be solved by the cooperation of compiler and operating systems to define a working area containing only global variables that may be accessed by both caller and callee. This probably involves the use of base registers set to provide a base for the caller, a base for the callee, and a base for the common area. However, if the base registers are part of the general register set this will not provide inviolable protection.

When the address ranges of caller and callee are disjoint because they are using nonmanipulatable base or segment registers that define separate space, the problem becomes one of how to get addressability of global variables to the callee rather than one of how to constrict addressability. There are various methods for doing this that are similar to the above. Protection depends on the control of access to base registers. The global variables are defined in an isolated segment, the segment address is made available to the callee, perhaps on a stack, and the callee (inner procedure) addresses the global variables through the global linkage segment register.

Some language theorists have commented that the concept of global variables and the scope of name rules are more trouble than they are worth because of the protection and addressing issues that they raise.

## QUESTIONS

1. What is the basic subroutine linkage instruction?

2. What spaces in the architecture of a processor have been used to hold the return address?

3. What is a linkage stack?

4. What stack operation can be used to effect a simple parameterless CALL?

5. How can POP be used to support RETURN?

6. What is an activation record?

7. How can the concept of activation record be supported by the use of a linkage stack?

# Chapter 9

# Interrupt Mechanisms and Control States

## 1. INTERRUPT MECHANISMS

Chapter 1 introduced the concept of an interrupt. The interrupt is a mechanism for diverting the attention of a processor when a particular event occurs. The architecture of an interrupt system involves the definition of the set of events that can cause an interruption, the definition of places in the processor where the information about the cause of an interrupt can be stored, and the specific activities that occur as a result of the occurrence of an interrupt.

Interrupts are used to support concepts of simultaneity and multiprogramming. For example, the mechanism permits I/O to proceed in parallel with computation and provides a way to synchronously report the completion of I/O functions to the processor. Interrupts also provide a way to limit the amount of continuous time a program can run so that computer time may be apportioned among active terminal users in such a way that each terminal user is made to feel he has continuous attention from the computer.

Interrupts are also used when some malfunction of the processor is recognized or when some improper condition of function or data occurs as a result of the execution of an instruction or the attempt to execute an instruction. These types of interrupt are sometimes called *exceptions* or *traps*.

## 1.1. Introduction to Interrupt Response

An essential decision in the architecture of an interrupt mechanism lies in the extent to which interrupts of a particular type are explicitly reported to the system. The most basic, and never actually encountered, interrupt architecture has just one type of interrupt that reports the occurrence of all events. Such a system might use specialized memory locations as follows:

| Mem Location | Use |
| --- | --- |
| 00001 | Address of first instruction to be executed after an interrupt. |
| 00002 | Contents of control counter at time of interrupt placed here. |
| 00003 | Interrupt status word placed here. |

At the end of the execution of an instruction, an interrupt signal is raised, causing the contents of the instruction counter to be placed in 00002. The contents of the instruction counter is then the next instruction to be executed in the flow of the interrupted program. Following this, the contents of 00001 are forced into the instruction counter. This causes a transfer to the instruction whose address is in 00001. This instruction is the first instruction of a *first-level interrupt handler*. The first-level interrupt handler accesses the *interrupt status word* in location 00003 to determine the cause of the interrupt. This status word provides information about the nature of the interrupt. A hypothetical 16-bit status word might look as follows:

| | |
| --- | --- |
| Bit 0 | Arithmetic overflow |
| Bit 1 | Addressing exception |
| Bit 2 | Illegal instruction execution |
| Bit 3 | Incorrect instruction |
| Bit 4 | Data parity error |
| Bit 5 | Supervisor call |
| Bit 6 | I/O error |
| Bit 7 | I/O completion |
| Bits 8–15 | Address of I/O device or not used |

The first-level handler inspects the bit pattern of this status word to determine the cause of the interrupt. If bit 0 is on, the interrupt occurred because a value exceeding the range of the arithmetic register was developed as a result of an arithmetic operation. Bit 1 indicates that the interrupt resulted because the program attempted to reference a memory location beyond its memory addressing range limits. Bit 2 indicates that an attempt was made to execute a privileged instruction by a program not in control state. Bit 3 indicates that the hardware could not recognize a bit value for an operation code. Bit 4 indicates that there is an improper number of bits brought from memory. The concept of *parity* provides for an extra bit to accompany each word in the system. This bit is 1, for example, if the number of 1 bits in the word is even, 0 if the number of 1 bits is odd. A parity error occurs when the bit count is not consistent with the parity indicator. Bit 5 indicates that an instruction asking for the processor to be put in control state was issued. Bit 6 indicates that an error occurred during an attempt to do an I/O operation, bit 7 indicates that an I/O operation was successfully completed. Bits 8–15 contain the address of the device if the interrupt was caused by I/O; they are empty if the interrupt was not caused by I/O.

The first-level interrupt handler determines the cause of the interrupt by masking out bit values until it finds a non-zero register. When the cause of the interrupt is determined, the first-level interrupt handler transfers to a program that is to handle the condition that caused the interruption. When this processing is complete, the original, interrupted, program may be returned to the processor. Its next instruction is located in memory location 00002.

## 1.2.  Interrupt Inhibition

While first-level interrupt analysis is being done, and while the condition indicated by the interrupt is being responded to, it is necessary that another interrupt does not occur. Another interrupt in this interval would put the instruction counter in memory location 00002, overlaying the address placed there by the first interrupt and destroying the ability to return to the initially interrupted program. We must have the concepts of *interrupt enable* and *interrupt inhibit*. These define the ability of the processor to accept another interrupt. There are many variations in the way that these concepts are represented in processors. One basic way is to provide INTERRUPT INHIBIT and ENABLE instructions. An INTERRUPT INHIBIT instruction is the first instruction executed by the first-level interrupt handler. The INTERRUPT ENABLE is executed when it is safe to allow another interrupt.

Exactly when it is safe to allow another interrupt may depend on the programming of the first-level interrupt handler and its condition handling successors. If the contents of memory location 00002 and 00003 are stored in some other locations in memory, then when they are removed from the lower memory locations, it is safe to allow another interrupt. The first-level interrupt handler may, in any case, be required to store the registers of the interrupted program in order to define a resumption point. The software structure that provides room for the register values at the time of interruption may also be used to hold the contents of the location counter placed in 00002 and the status word in 00003.

## 2.  VARIATIONS IN INTERRUPT ARCHITECTURE

The basic interrupt mechanism involving three locations in lower memory, a first-level interrupt handler, and INTERRUPT INHIBIT and ENABLE instructions is useful only as an introduction to the phenomenon of interrupt handling. Architectures vary quite widely in the definition of their interrupt structures. Among the variations are the following:

1.  The specific number of interruption causes treated as unique events. For example, many processors have multiple interrupt status words, one used for reporting I/O events, one used for reporting processor events like arithmetic overflow, and so on.

2. The specific means for enabling and disabling interrupts, and the specificity with which certain types of interrupt may be enabled or inhibited. For example, some processors may inhibit I/O interrupts while enabling processor interrupts.

3. The specific use of lower memory locations and/or special registers for reporting information about an interrupt or for representing the enabling or disabling of interrupts.

4. The extent to which the concept of interrupt priority is associated with the interrupt mechanism and the particular ways that priority is used to sequence interrupt handling.

5. The amount of program data that is actually automatically stored in memory at the time of an interrupt and the amount of data that is taken from memory and put in processor registers. For example, some architectures store not only the instruction counter but all or some of the operand registers and selected control registers in memory, or they store specialized registers automatically at the time of an interrupt.

6. The exact times when an interrupt can be accepted. Many systems will accept interrupts only at the completion of an instruction, others will accept them during instruction execution. Others will accept some interrupts during instruction execution but others only after an instruction is complete. The time of the acceptance of the interrupt depends on whether the interrupt is caused by some part of the system that has no relationship to the program executing or whether the executing program caused the interrupt. An interrupt caused by an executing program is not necessarily an error; it may also be the result of some legitimate reason for generating a response to a particular condition. Such a condition might occur in a paging machine when it is found that a referenced page is not in memory and the operating system must fetch the page to memory.

### 3. GENERAL DESCRIPTION OF S/370 INTERRUPT ARCHITECTURE

Most processors have interrupt structures that are considerably more elaborate than what has been described so far. More elaborate interrupt structures involve the notions of classes and levels of interrupt. The basic scheme is to provide additional and more specific interruptions so that interrupt analysis can be reduced and the interval of time required for the inhibition of interrupts is minimized. We will use the S/370 as a model for describing a somewhat more elaborate interrupt architecture than we used for our introduction.

### 3.1. Interrupt Classes

The S/370 architecture recognizes six classes of interrupt: *external, I/O, machine-check, program, restart, and supervisor call.*

The class of external interrupt fundamentally contains specific interrupts that have to do with an operator pressing a key on a console that will result in an interrupt; with some error in the clocks or timers associated with the system; with action on a set of external signal lines; with actions between two processors in a multiprocessor configuration; or, as a result of a malfunction, with an alert signal sent by another processor in a multiprocessor configuration. It is the responsibility of interrupt handling analysis programs to determine which of the specific kinds of external interrupt signal has been generated.

A machine-check interruption is generated as a result of an equipment malfunction. These interruptions include parity errors. (The S/370 architecture carries extended parity information so that some recovery can be undertaken on a parity error. The parity bits are called *error checking and correction codes* and provide for correction of single bit errors and determination of double bit errors on transfers from memory.)

A program interruption is a result of improper operation codes; addressing beyond the range of memory locations configured with the processor; when an arithmetic instruction is applied to a wrong data coding; when an attempt is made to divide by zero; when there is decimal overflow, or floating point overflow, or underflow of some other type; when an address lies beyond the legitimate addressing range of the program; when a privileged instruction is executed by a program not in supervisor state; and so on. In general these interruptions are concerned with the running program and result in conditions caused by the operation codes or addresses of the running program. The interrupt analysis function must determine which specific interrupt has occurred within the class of program interrupt.

An I/O interrupt provides a means of coordinating processor and I/O device activity. Interrupts may occur as a result of a busy device, a malfunctioning device, or successful completion. The interrupt analysis routine must participate in determining the cause of the interrupt and the exact device that caused the interrupt.

The restart interrupt is provided to enable an operator to invoke a particular program from the console. It is unique in that it cannot be disabled and will start execution of a processor when the processor is in a stopped state.

The supervisor call interrupt occurs when a program issues an instruction, SUPERVISOR CALL (SVC), requesting that a program that is part of the operating system run. This does not automatically place the system in control or supervisor state, as we shall shortly see.

The specific details of processor reaction to each of these interrupts differ from interrupt to interrupt and are somewhat model dependent. We undertake a general description of interrupt responses that will necessarily omit some details.

## 3.2.  The S/370 Program Status Word

In order to describe the behavior of the interrupt mechanisms we must discuss an S/370 structure called the *program status word* (*PSW*). The PSW is a control register that provides the functions of a location counter and contains bits that describe the interrupt and machine status of the processor. The immediate action of an interrupt is to place the contents of the PSW in an assigned location in memory and to place in the PSW a new PSW word that causes the execution of an interrupt analysis function.

The PSW is a 64-bit control register organized into major functions such as *system mask, PSW key, condition code,* and *program mask.* The bit layout of a PSW is as follows:

| Bits | Use |
| --- | --- |
| 0–7 | System Mask. |
| 1 | Determines whether processor will accept program-event recording interrupts. (We have not described these interrupts in text.) |
| 5 | Indicates whether or not virtual address translation should occur. |
| 6 | Determines whether the processor will accept an I/O interrupt. An associated control register, called the channel mask register, determines which specific channels (pathways to devices) can cause an interrupt. |
| 7 | Determines whether the processor will accept external interrupts. |
| 8–11 | Protection key. The architecture uses the concept of a protection key to control the addressing of programs. In effect, a value is assigned to be a key for a particular area of memory. When a program address is developed, the key of the area is compared with the key in the PSW. If they match the reference is permitted. If not, an interrupt occurs. |
| 12 | Control Mode. The architecture may run as an S/370 or an S/360. This bit determines the exact form of the PSW and the use of various control registers. The format we are describing is S/370, extended control mode. |
| 13 | Determines if the processor will accept machine-check interrupts. There is an associated control register that describes more complete enabling of machine-check interrupts. |
| 14 | Indicates that the processor is in a wait state. A wait state means that the processor is not executing instructions but may accept interrupts. |
| 15 | Indicates whether the processor is in supervisor or problem program state. Attempts to execute privileged instructions in problem state will cause a program interruption. |
| 18–19 | Condition Code. These 2 bits contain values developed by the operation of certain arithmetic or comparison instructions. As we saw in Chapter 7, they are used by branching instructions to test for the existence of a certain condition as the basis for a BRANCH. |

| | |
|---|---|
| 20–23 | Program Mask. Determine whether or not various arithmetic operation conditions will result in interrupts. |
| 20 | Fixed Point Overflow. |
| 21 | Decimal Overflow. |
| 22 | Exponent Underflow. |
| 23 | Significance. |
| 40–63 | Instruction Address. |

We see that the PSW is used as the instrument for indicating inhibition or enablement of various interrupt classes and controlling addressing and as an instruction counter and a control or problem state indicator. Thus for any program the interruptions that can be accepted, the operation codes it can use, as well as the memory locations it can address are defined by values in the PSW.

A full description of interrupt status is not provided by the PSW since support of a channel mask control register is needed to further define from what I/O sources an interrupt will be accepted. A machine-check control register is needed to define what machine-check sources will be accepted.

The immediate response to an interrupt is to place the PSW in a location in memory and to bring from another location in memory a new PSW to control the processor. The new PSW provides the address of the first instruction to be executed after the interrupt. The bits in the systems mask, program key, and program mask sections change the status of the machine as regards control state and interrupts and addressing permitted.

### 3.3. Organization of Lower Memory

The first 512 locations of lower memory are used for special control purposes in some versions of this architecture. We show only those locations of interest in this discussion of the interrupt architecture.

| Decimal Address | Use |
|---|---|
| 0000–0007 | PSW for Restart Interrupts. This value is placed in the PSW register when a restart interrupt occurs. It is called the *restart new PSW*. The phrase "New PSW" is generally used to refer to the PSW from memory being placed in the PSW register. Eight bytes are required for the 64-bit PSW value. |
| 0008–0015 | Place to put PSW of interrupted program when a restart interrupt occurs. Generally these are called *old PSWs*. |
| 0024–0031 | External interrupt old PSW. |
| 0032–0039 | Supervisor call old PSW. |
| 0040–0047 | Program interrupt old PSW. |

| | |
|---|---|
| 0048–0055 | Machine-check old PSW. |
| 0056–0063 | I/O old PSW. |
| 0064 0071 | Channel Status Word. This value is set as a result of an I/O operation. It indicates the status of an I/O operation. It is this value that is inspected by an I/O interrupt analysis routine to determine if I/O was successful or if some error occurred. |
| 0080–0083 | Interval Timer. A clock that is constantly lowered when interval timing is enabled. It may cause various kinds of interrupts. |
| 0088–0095 | External New PSW. The value to be placed in the PSW register when an external interrupt occurs. |
| 0096–0103 | Supervisor call new PSW. |
| 0104–0111 | Program interrupt new PSW. |
| 0112–0119 | Machine-check new PSW. |
| 0120–0127 | I/O new PSW. |
| 0132–0133 | Holds the address in the processor associated with interrupts of the external call, malfunction alert type. |
| 0134–0135 | External Interruption Status Word. Holds information necessary for analysis of an external interruption. |
| 0136–0139 | Supervisor Call Interruption Identification Data. Holds information necessary for analysis of supervisor call interruption. |
| 0140–0143 | Program Interrupt Data. |
| 0144–0147 | Translation Exception Address. Holds an address that caused an interrupt associated with virtual addressing. |
| 0168–0171 | Channel Identification Data. Identifies the source of an I/O interrupt. |
| 0185–0187 | Provides additional I/O addressing information. |
| 0187–0511 | Provides additional space to support necessary data to analyze and recover from machine-check interrupts. |

From the list of locations above we need understand only that each class of interrupt has a special location in which to store the old PSW and from which to get the new PSW, and additional locations that, for each class of interrupt, provide more information about the exact nature of the interruption. Remember that the inhibiting or enabling of various interrupt classes is done by the PSW in association with a set of control registers that provide more information for machine-check and I/O interruptions. The exchange of PSWs provides for changing the interrupt enable status and the status of addressing and privilege for the program running under the PSW in the PSW register.

### 3.4.  Interrupt Response

The exact details of interrupt response differ from class to class. In general, an interrupt causes the storing of the old PSW in the location assigned to the interrupt class, the retrieval of a new PSW from the location assigned to the interrupt class, and an inspection of status words associated with the interrupt class.

There is a fixed priority relationship between the classes of interrupt. If all interrupts are enabled and simultaneous interrupt requests occur, then the interrupts will be serviced in a predetermined order. A machine-check interrupt is always serviced first. After a machine-check interrupt the priority order is supervisor call, program, external, I/O, and restart. Rather complex strings of interrupts can be successfully handled. Priority is also applied to the concept of disabling interrupts. A processor that is inhibiting an interrupt of a lower priority may accept an interrupt of a higher priority, even if the lower priority interrupt is still being processed.

This architecture is a considerable elaboration of the simple interrupt scheme used as an example. It provides for a reduced interrupt analysis function since the class of interrupt is always known as a result of the PSW that is established in the PSW register, and it provides for a priority of interrupt among interrupt classes. The concept of the single interrupt status word is generalized to provide a set of interrupt status words providing information about the reason for each class of interrupt, and a quite flexible inhibition scheme determines when it is safe to allow other interrupts. Interrupt enable status change may also be accomplished by the use of privileged instructions that can change the bits in the interrupt masks.

## 4.   OTHER INTERRUPT SCHEMES

We have mentioned that interrupt architectures may vary widely. This section describes some features of the interrupt architectures for processors other than the S/370. We describe only those features that represent concepts that differ from the S/370 described above.

The CRAY-1 undertakes to incorporate the necessity for storing operand and certain control registers for an interrupted program in the interrupt architecture. The goal is to reduce the amount of time it takes to clear the registers of data about an interrupted program and fill them with data about a replacement program. This is particularly important to this architecture because of the large number of operand registers that exist in the architecture.

The architecture has the concept of an *exchange package*. An exchange package is a 16-word package that consists of:

1.   All eight scalar (S) registers.
2.   All eight address (A) registers.
3.   The contents of the location counter.
4.   The base address and address limit registers.

5.   A set of bits that indicates the cause of the interruption. Interrupts may be caused by a memory error, I/O, floating point arithmetic errors, or addressing errors.

Thus the exchange package contains both register values and source of interrupt information. Notice that not all of the registers of the CRAY-1 are included in the exchange package. The operating system stores the vector (V), address save (B), and operand save (T) registers in a software defined area in memory.

When an interrupt occurs, instruction execution is suppressed and all memory operations are allowed to complete. As a result of the interrupt, the contents of the exchange package are formed and placed in positions in lower memory and a replacement exchange packet fills the registers. This exchange makes it unnecessary to save the S and A registers by programming.

Some architectures make a distinction between interrupts and *exceptions*. The word "interrupt" is used solely to designate the reporting of external, asynchronous events to the processor. The word "exception" is used to describe changes in processor execution flow as a result of conditions resulting from instruction execution. Among the processors that make a distinction of this type are the IBM Series/1 and the VAX-11.

In the VAX-11 architecture the exception is handled as an event occurring within the context of the running program, and condition handling is done in a manner somewhat like calling a specialized subroutine. The VAX-11 provides exception enabling or inhibition bits in its PSW for some exceptions relating to arithmetic operations.

Interrupts are associated with external events and may not be inhibited. A priority scheme is associated with interrupts so that each interrupt request has a particular priority. Priority levels are assigned to operating system functions, hardware interrupts, the systems clock, and processor error conditions.

The Series/1 and the IBM 8100 associate interrupt priority levels with independent sets of registers. In the IBM 8100, for example, there are eight priority levels, each of which is associated with a set of operand and control registers used by a particular program running at that priority level. Each interrupt level has both a problem state and a control state. The intent of architectures of this type is to permit real-time functions that are very critical to associate their devices with higher priority levels so that they are accepted and serviced before devices of lower priority.

## 5.  CONTROL STATES

We have been making reference to the existence of a control state, or supervisor state, in the architectures of many processors. This control state has the following features:

1.  It permits the privileged instructions to be executed.

2.  It permits control registers to be altered in ways only available to programs running in this state. It may accomplish this by permitting the execution of instructions that can manipulate these control registers.

3.  It provides for interaction with the I/O devices of the system. In many processors the execution of I/O instructions is a privileged operation. The restriction on the use of I/O instructions is to facilitate coordination of programs that share the same I/O devices. In an environment of multiprogramming it is possible that programming and data for more than one active program will be held on the same storage device. In order to properly sequence references to a shared device, as well as to relieve applications programs from the complexities of I/O programming, issuance of I/O instructions is centralized within the operating system in control state.

4.  The addressing constraints applied to programs not in this state may be modified. This modification may include an ability to modify protection keys so that the operating system running in control state may address all of memory. The special control memory locations, in particular, may be addressable only in control state.

Control state is always indicated by some bit or flag in some processor control register. In S/370 architecture it is indicated in the PSW, as we have seen. In many architectures, entry to the control state is automatic at the time of the occurrence of an interrupt. In some systems, bits in the new PSW will determine whether the interrupt handling routine will operate in control state.

In some architectures, among them the VAX-11, the concept of control state is enlarged to include a number of levels of privilege and control. The VAX-11 contains four *access modes* that define what a program running in a particular state may do. These four access modes accord with a layered software structure of an operating system in which there is a kernel state, an executive state, a supervisor state, and a user state.

These states are reflected in a *processor status word*. The user state is for applications programs with the most restricted privileges. Each other mode accords with a concept of enlarged privileges associated with different operating system functions. Thus command language interpreters run in supervisor state, various systems services run in executive state, and basic control over all memory and all instructions is available in kernel state. There are instructions that change the mode from one level of control to another. There are also instructions that protect a program running at a more privileged level from making errors as a result of some error in a program it is serving which is running at a higher level.

The reader will see a similarity between the levels of control described in this chapter and the notion of protection rings discussed in Chapter 5. There are architects who feel that control states are not necessary if addressing control and constraint are nonviolable and that control of addressability is sufficient to protect the control mechanisms of a machine if the only way to produce code is through a compiler that refuses to generate privileged instructions.

## QUESTIONS

1. What is the purpose of an interrupt?

2. How does a processor respond to an interrupt?

3. What is the concept of interrupt inhibition?

4. What various kinds of interrupt class may exist?

5. What is a PSW? In general? In the IBM/370 architecture?

6. How do lower memory address locations participate in interrupt definition?

7. Distinguish between an interrupt and an exception.

8. What kinds of things can be done in control state that cannot be done in noncontrol state?

9. Discuss why more than one level of control state might be useful.

# Chapter 10

# Virtual Memory

## 1. CONCEPTS OF VIRTUAL MEMORY

In this chapter we explore the architectural concepts that are relevant to various notions of virtual memory. The underlying concept of virtual memory is that the addresses that are available to a program have no direct relationship to physical locations on a machine. There is no way to directly infer the actual memory locations represented by an address from an inspection of the address. This concept has two aspects of varying importance to different schools of architecture and organization.

One aspect of the disassociation of an address from a memory location is that the range of addresses that may be formed for a program is larger than the available real physical memory locations in any configuration of a machine. Thus a program can be written that is a million locations long and will run on a configuration that has only 250,000 locations of memory. The intent is to free the programmer or compiler from concerns about fitting programs into available memory locations. This notion of virtual memory, concerned with the enlargement of a program's address space, tends to conform to the view that memory is a linear, one-dimensional set of locations ranging from 0 to the largest address that can be formed by address arithmetic. Operating systems that support this concept of virtual memory may define one or more virtual memories in the system. If one virtual memory is defined then all programs are allocated space in the single enlarged virtual memory as if they were in real memory. The total number of allocated addresses, however, is the size of the virtual rather than the real memory. If multiple virtual memories are provided by the operating system, then each program running in a multiprogramming or time-sharing mix receives an address space the size of the largest addressable range defined by address arithmetic.

The linear concept of virtual memory is preoccupied with the enlargement of addressable space and with relocation. As we see in this chapter, a fundamental function of the definition of a hardware mechanism to support virtual memory is the ability to dynamically relocate portions of programs and to provide for more intensive use of memory locations by avoiding fragmentation. The portions of a program that appear to be contiguous in memory may be scattered throughout the memory in fixed size blocks of locations called *pages* This ability to scatter program elements provides an ability to allocate locations without regard to their contiguity in physical memory and makes it easier to avoid the occurrence of unusable areas of memory as programs enter and leave the multiprogramming mix. A small physical location space can be used for large virtual address space because of the dynamic reassignment of virtual addresses to physical addresses as the program runs.

Another notion of virtual memory, however, presents the idea of memory as a two-dimensional array and the contents of a memory as a set of named logical structures, each one of which is addressed by a name and a relative position within the name. The concern of this concept of virtual memory is not relocatability but addressability. The concept is implied by our discussion of segmentation in Chapter 5 and discussed again in the section on capabilities in this chapter. This two-dimensional view of virtual memory is not concerned primarily with the relative size of virtual address space as a linear array and physical locations as a linear area. It views the contents of memory at any one time as being a collection of uniquely named *objects* or *segments* whose ordering and relationship to each other cannot be inferred from their virtual addresses. In a linear virtual address the virtual address 570009 is assumed to be sequentially higher than 456011. The addresses may not be mapped into sequential memory locations but the virtual address space is considered sequential by virtual address. In a two-dimensional view of virtual memory, 570009 might be interpreted to mean segment 570, relative location 009. 456011 would be interpreted to mean segment 456, relative location 011. The higher sequential value of segment 570 relative to 456 has no significance, and the two values are treated as segment names with no ordering relation in virtual memory implied.

In a two-dimensional system, a program, a data area, a working area, even an area on disk, may be considered to live in its own private virtual address space which is, in effect, a small virtual memory. One distinction between the linear and the object concept of virtual memory is that, in systems built primarily around the linear concept, we tend to talk about *cross virtual memory* communication when a program running in one virtual memory wishes to talk to a program running in another. Contemporary linear virtual address operating systems are just starting to support the idea of cross virtual memory communication. In two-dimensional systems we talk about object or segment access and object or segment linkage and assume that it is common and going to occur between elements of a program as a natural part of running the program. Thus, programs are viewed as having a great deal of internal structure and a set of external relationships with other objects.

In effect, the two-dimensional concept of virtual memory addresses relationships between logical segments or logical objects without regard for physical relocatability. These schemes are segment-oriented rather than page-oriented and may or may not include notions of enlarged addressing spaces. In fact, many proponents of segmentation-based concepts of virtual memory believe that the address space of a particular program should not be enlarged, as they are in linear virtual memory support, but constrained to encourage modularity of program definition for ease of debugging and modification.

Those who feel that a fundamental notion of virtual memory must include concepts of address space enlargement over physical locations look upon two-dimensional concepts not as virtual memory concepts at all, but as merely protection mechanisms.

Strictly speaking, a programmer or compiler may ignore the existence of the idea of linear virtual memory. Systems programmers, operators, and others, however, are aware of the concept. The existence of the idea of virtual memory may intrude on a programmer desiring to write excellent and optimized code. However, ideas of linear virtual memory do not normally involve any instruction by instruction decisions on the part of the programmer or compiler.

Segment-oriented virtual memory as an addressing concept, however, becomes part of a fundamental concept of programming and relates to notions of module definition very closely. Some see this as a disadvantage in that a programmer must plan module and segment structures carefully. Others see this as a tremendously important programming style—not a burden imposed by segmentation concepts but a reform of programming practice. The burden of planned overlays and segment loads, characteristic of programming in machines without any concept of virtual memory, may be removed from a programmer by mechanisms in the hardware or operating system that automatically bring referenced segments or objects into real memory as they are required.

## 2. LINEAR VIRTUAL MEMORY

The simplest concept of virtual memory provides an extended linear address space with the intention of providing the programmer with an address space that is significantly larger than the number of physical locations available on the target machine. This illusion of larger space is created by the joint efforts of processor hardware and features of a virtual memory operating system. The hardware and programs of the operating system combine so that the addresses referenced by a running program are brought into memory in such a way that the programmer has the conceptual ability to treat the memory as if all that he wished were in it actually were at all times.

Since the hardware structures that support the relocatability of virtual memory are intended to be invisible to a compiler or programmer, they are really more organizational than architectural in nature. However, virtual memory has become such a part of the general culture of computing that we describe the supporting mechanisms as architectural mechanisms. There are actually a large number of professionals who are very aware of virtual memory and must make operating system parameter decisions about what parts of programs must map into physical locations identical to the virtual addresses (VIRTUAL-REAL), what parts can be relocated but not dynamically moved (VIRTUAL-FIXED), and what parts can be freely relocated. In addition, there is a necessity for determining the permissible ratios of virtual to physical address spaces, the number of virtual memories that can be defined, and so on. All of this is part of the performance planning of a configuration.

## 2.1. Basic Mechanisms

Two fundamental notions are required to understand linear virtual memory: We need the idea of a *page* and the idea of a *block* of real memory. A page is generally an addressable area of fixed size. Frequently used page sizes are 2K and 4K bytes. This means that (for a 4K page) each page is addressed from 0 to 4K-1 (0 to 4095) bytes. A page is the logical portion of a program associated with real physical memory locations. Page addresses are virtual addresses.

A block of memory is a set of physical locations in memory to which a page is assigned. It is the address of these physical locations that must be derived from any *address translation mechanism* that translates from program virtual addresses to real physical locations. The translation mechanism is fundamentally a table that relates program virtual addresses to physical locations in terms of individual page/block relationships.

The mapping of program virtual addresses into physical locations is commonly accomplished by a *page table*. A page table is an area of memory that holds a list of pages and associated blocks. Access to the page table in memory is indicated by the value in a *page table origin register* that points to the area in memory where the page table begins.

The table below shows a page table organized for eight entries. The page table has been placed to begin at location 1000 (by the operating system). The page table origin register has been set to 1000, indicating the start of the page table. The table below assumes byte addressing and 4 bytes per entry. Each entry contains the starting location of a 4K memory block that currently holds pages representing the virtual memory addresses indicated by the column Represents Program.

Page Table Origin Register: 10000

| Page Table | Memory Block | Represents Program |
|---|---|---|
| 1000 | 8192 | 0 to 4K-1 |
| 1004 | 32768 | 4K to 8K-1 |
| 1008 | 16384 | 8K to 12K-1 |
| 1012 | 65536 | 12K to 16K-1 |
| 1016 | 98304 | 16K to 20K-1 |
| 1020 | 81902 | 20K to 24K-1 |
| 1024 | 163804 | 24K to 28K-1 |
| 1028 | 4096 | 28K to 32K-1 |

The table shown above indicates that a program address developed by usual combinations of addresses in instructions, index registers, and base registers may range from 0 to 32K-1. The program is 32K locations long. These 32K locations are divided into eight pages. Each of the eight pages is individually allocated to blocks of memory as indicated on the table. Thus, any address formed by address arithmetic in the range 16K to 20K-1 will be found in the block of physical memory starting at location 98304. A program address of 20K, however, will be found in physical memory in the block beginning at 81902. The exact method for finding appropriate entries on a page table of this type is described in the next section.

This noncontiguous and nonsequential dispersion of the program through memory space allows the operating system to make maximum use of memory space, reducing the occurrence of unusable memory fragments and the need for the intermittent rearrangement of memory to fit in programs as the population of programs running on the machine changes.

When a program starts to run on the processor the page table origin register is loaded with a value indicating the location of its page table. This function is performed by the operating system in control state because this register is not made available to programs running in problem state. Since the page table defines the addresses in a memory that a program may reference, its manipulation and accessibility are generally secured by control state and a privileged instruction.

Before the program actually starts to run, it must be brought into memory and its pages assigned to memory blocks. The operating system will determine what memory blocks may be made available to pages of the program and assign and load each page individually. When not in memory the program pages are on a disk or drum. The operating system must know the location on these devices of each page of a program in order to fetch it into memory. The parts of an operating system concerned with these activities are the *main storage manager* and the *auxiliary storage manager*. The main storage manager determines what blocks to load pages into. The auxiliary storage manager finds the pages on drum or disk. The auxiliary storage manager relies on the support of I/O elements in the operating system. Page addresses on disk or drum must be translated into I/O orders and device addresses (discussed in Chapter 11) in order to cause the movement of pages from drum or disk to main memory.

## 2.2. Memory Space Management

Operating systems differ enormously in the sophistication of algorithms relevant to finding free blocks in memory and handling pages on disk or drum. One difference between approaches to paging is whether the operating system uses *swapping* or *demand paging*. In pure swapping all of the pages of a program are loaded into memory blocks when the program begins. Thus, in order to start, all of the program must be contained in memory. In pure demand paging a program is started when some minimum space is available and additional pages are brought into memory only when they are referenced from pages already in memory. The notion of demand paging is that a program displays a large amount of *locality of reference*. That is, for long periods of time a program tends to address in a particular area of data or instructions.

The amount of space required for a program to perform well, not to be intolerably delayed by having to acquire pages from disk, can be determined by observation of the *working set*. A working set of pages is the set of pages most recently referenced by a running program. Swapping systems may be *working set guarantee systems* in that they swap into memory at the start of a program those pages the program referenced most recently (during the time of its last execution). We assume a pattern of rotating execution in a multiprogramming environment where a program runs for a period, is suspended, loses its place in memory, and, when restarted, swaps in its working set. The intent is to minimize paging during operation of a program while minimizing the amount of space used by a program while it is running. In this way a maximum number of programs may be placed into a mix and the system can be intensively utilized.

A demand paging system may be a *working set size guarantee* system where the requirement for starting a program is that enough memory blocks may be made available to hold the number of pages used during the last execution. The actual pages, however, are not swapped in. This approach minimizes page movement if the *neighborhood* of a program changes from one execution period to another. Certain features of I/O, however, may make swapping efficient even if a relatively large change in addressing patterns occurs after a reactivation. A large number of pages addressed by a single I/O command may be brought into memory in considerably less time than were each page individually moved.

The decisions about swapping, demand paging, and other related details are operating system design decisions and not computer architecture decisions up to this point in the development of computers. Factors relating to operating systems design in this area are discussed in Lorin & Dietel (noted in the bibliography) and in many other places. Hardware support for paging is currently limited to mechanisms that determine whether or not a page is actually in a block and that respond to a page not in a block being referenced and to mechanisms that help determine the working set and make the process of address translation somewhat faster.

## 2.3.   Address Formation in Paging Systems

When the program starts to run its addresses are formed in the usual way using instruction address fields, index registers, displacement registers, and so on. This address formation, however, does not lead to a machine location but to a fully formed virtual address in the area of the program. A machine address is formed when the hardware uses the virtual address in connection with the page table.

To find a physical location from a virtual address the interpretation of an address splits it into two areas, a page address portion and a relative position in page portion. The upper bits of the virtual address, the page address portion, is added to the page table origin register to form an index into the page table. The contents of the selected location in the page table represent the starting location of a 4K block of memory. The lower bits of the virtual address are added to the contents of the page table to form an actual machine address.

The process of address formation, using the page table shown above, is demonstrated below:

### Virtual Address Is Formed

| | |
|---|---|
| Address in instruction word: | 00578 |
| Index register: | 10433 |
| Base register: | 14567 |
| Virtual address: | 25578 |

### Virtual Address Breakdown

Virtual address: 25578 (Decimal) = 0110001111101010 (Binary)
Page size 4K, 12 bits to reference in a page: 001111101010 (Binary)
Page relative address: 1002 (Decimal)
Page address portion: 4 bits, 0110 (Binary)
Page address designator: 6 (Decimal)

### Access to Page Table

Page table origin register: 1000
Page address designator: 6
Adjustment for 4-byte entries: 6*4=24
Indexed entry on page table: 1024
Contents of indexed entry: 163804
Offset for page relative portion: 163804+1002
Final physical memory location: 164806

## 2.4. Page Faults

For as long as virtual addresses translate into pages in memory nothing more need be done. It is expected, however, that a running program is greater in size than the 32K representable in the example page table and that a virtual address may be formed that references a page not assigned to a memory block. For example, a virtual address of 42618 would not find an entry on the page table. If the page table were bigger and held an entry for all possible virtual addresses, the discovery of a non-memory resident page would be made by finding some entry in the table that indicated that no memory assignment had been made for this address. Using the simple indexing scheme we have been discussing it is in fact necessary that the page table be long enough to represent all possible page addresses. Growth and contraction of page tables may be accomplished with an elaboration of the mechanism described later in Section 2.6.

When the page table search determines that the referenced page has not been assigned to a block, steps must be taken to find the referenced page on disk or drum, allocate a memory block to it, and bring the page into the assigned memory block. The hardware event that starts this process is a *page fault interrupt* or *page fault exception*, which signals the processor that a referenced page is not in memory. This signal causes the execution of the operating system interrupt analysis routine. The program causing the page fault interrupt is suspended and the analysis routine determines that the interruption was caused by a paging fault. The interrupt analysis routine then asks for the execution of the auxiliary storage manager and the memory manager.

The function of the auxiliary storage manager is to provide the address on disk where the referenced page is stored. The function of the memory manager is to provide an address in memory to which the page may be read. These procedures may be relatively straightforward if the memory manager can find empty space for the new page. If it cannot then space must be taken from a page currently in storage. An attempt is made to identify the page with the smallest probability of being referenced again that has not been changed. The inference of probability of near reference is usually based on keeping a record of last references and predicting that pages least recently used have fallen out of an addressing neighborhood and have a low probability of being referenced again. An additional bias for selecting a page that was not changed while it was in memory exists in most operating systems. This eliminates the necessity for saving it on the disk before using its space.

Various forms of architectural and organizational support are provided for determining how recently a reference was made and whether a page was changed. One approach is to take bits in the page table to indicate the status of a page. An indication of reference and change is part of the entry for a page. While this does not approximate least recently used very closely, at least there is an indication of those pages not referenced or changed that are candidates for being overlayed in their block. More elaborate organizational schemes involve the representation of a set of or all of the pages in either a set of shift registers or a matrix. Each time a page is referenced, numeric manipulation is done in the mechanism so that an ordered list of references is provided.

When the operating system has determined the location on disk and a space in which to read a page, it enters its I/O programs with a request to read the page from disk into memory. Since there will be some delay in the actual arrival of the page, the operating system then typically enters its dispatcher to task switch to another program. The program that made the reference to a non-resident page is marked non-dispatchable until the page arrives.

The page fault mechanism is basic to the mapping of expanded virtual address space onto smaller physical space. It provides the central means for the dynamic reuse of memory locations by starting the process that will reassign a block of memory locations to a different page.

## 2.5. Linear Virtual Memory and Protection

The basic linear model of virtual memory does in fact provide for an association of addressing and protection since it is impossible for a program to refer to a memory location that is not allocated to a page in its own virtual memory. Sharing memory locations is accomplished by pointing to common page tables, perhaps by recording the address of common page tables on a segment table. But within a virtual address space, protection must be achieved by means other than the addressing mechanism.

## 2.6. Segmentation in Linear Virtual Memories

We have already discussed the use of segment registers and some concepts of segmentation. In the context of linear virtual memory paging systems, the word segment takes on a meaning that is specific to the relocation mechanism rather than to addressing and protection. Segmentation may be thought of as an extension to the paging mechanism described so far. This mechanism has two problems: It is difficult to increase the size of a running program's virtual address space, and every memory reference involves an additional reference in order to use the page table to derive a machine location.

The solution to the problem of growing and expanding virtual memory size is addressed in 370 derivative architecture by the addition of an extra level of address formation called segments. The processor contains a *segment origin register* that points to the beginning of a *segment table*. The segment table contains the addresses of page tables. Addressing now involves using the upper bits of a derived virtual address as addends to the segment origin register. This indexes to the address of a page table. This address is placed in the page origin register. Middle bits of the address are added to index to a particular page table, and the lower bits of the address are used in the usual way to access a relative location within a page. The program can now grow and contract by adding or deleting page tables.

## 2.7. Associative Memory Assist

The addition of the segment table concept seems to make the memory addressing problem worse since it is now necessary to go to memory twice to form an address, the first to inspect the segment table, the second to inspect the page table. The solution to this problem is really an organizational not an architectural solution. It involves the provision in the system of a small amount of very fast memory that has the characteristic of being content addressable.

The content addressable memory is used to support paging systems by providing an entry for a subset of resident pages, the last pages referenced by the processor. After the formation of a virtual address, the associative memory is searched for the virtual address-machine address pair. When it is found, the machine address is formed and the reference to memory made. Only when a virtual address is not found in the associative memory is reference to the segment and page tables necessary. When a new page is referenced it is brought into the associative memory. Residence in that memory is by some hardware approximation of the least recently used algorithm.

Thus, at any one time the most recently referenced pages of a program are represented in the associative memory. In models of the 370 this memory is called the *table lookaside buffer*.

| Virtual Address | Real Address | Status |
|:---:|:---:|:---:|
| 00000 | 8192 | NOT REF |
| 04095 | 32768 | REF |
| 08191 | 16384 | REF |
| 16383 | 65536 | NOT REF |

The above represents a four-entry table lookaside buffer that maps virtual address pages and real memory blocks. It also contains a status indicator that indicates whether or not the page has been referenced recently. Since space in the table lookaside buffer is only for most recently referenced pages and the buffer may be small relative to program sizes, references are made to pages not in the buffer from time to time. This causes a reference to the page table such as was described in previous sections. The assignment of the newly referenced page is recorded in the table lookaside buffer. It must replace some mapping already in the buffer. The status bits provide an indication of those pages that have not been recently referenced. One algorithm, defined in hardware for the table lookaside buffer, sets all status bits to NOT REF when they have all been set to REF. In the ensuing interval some bits will be set back on to REF. These represent those pages used within the interval and consequently the pages most recently used. Entries whose status bits have not been turned back to REF are candidates for overlay for newly referenced pages.

In smaller systems the contents of the TLB may be searched in parallel for the appropriate virtual address/real address mapping. In larger systems it is common to use a hashing algorithm on the virtual address to select an entry in the TLB. This entry is then inspected for the proper virtual address. The hashing is done in order to avoid very expensive circuitry for the parallel search of large associative memories. Speed is important to this process since its intent is to minimize the time necessary to form real addresses from virtual addresses.

In systems that provide multiple virtual memories, such as the MVS operating system for the IBM 370 generic architectures, it is necessary to identify the specific virtual memory whose mapping is represented in a TLB entry. Thus a search of the TLB will not only look for an entry for the virtual address, it must check that the virtual address/real address mapping in the TLB represents a mapping of the correct virtual memory. This is necessary because each virtual memory has identical virtual memory addresses. Thus without the virtual memory identification check it would be impossible to know that the virtual address found in the TLB came from the virtual memory in which the program was running.

## 3. ALTERNATIVE ARCHITECTURAL CONCEPTS

The concept of virtual memory supported by the above provides for support of a concept of virtual memory as an enlarged linear array. Segmentation, in connection with this concept, is another level of relocation addressing. In fact the failure to find a page on a particular segment table may cause a search in some segment table represented as a linear successor by the virtual address.

In Chapter 5 we saw some other notions of segmentation having to do with the partitioning of programs into code and data portions, suggesting that program structure should be associated with addressability. Segments have no relationship to each other in the sense that one segment definition implies a certain part of a virtual address space, another segment implies another part, and so on. Segments are seen as independent logical elements participating in the definition of a two-dimensional address space always addressed by segment and relative position in segment. The logical unit of memory management is the segment. The movement of segments of arbitrary size may or may not be supported by a paging mechanism. The definition of virtual memory is oriented to addressability and protection rather than relocation.

One interesting recent development in this area concerns emerging concepts of the *object model of computing* implied by the term *capability based addressing*.

Capability or object management operating systems and architectures, of which the IBM System 38 and the INTEL 432 are examples, are motivated by a perception of computing that postulates that very large address spaces, created by linear virtual memory approaches, are not good features for an architecture. Many programming theorists believe that very large linear address spaces lead to poorly structured and difficult to debug programs as well as to possible problems in protection. A program in this context is seen as preferably having a relatively small directly addressable space and addressing out of this space through capabilities. A capability is a pointer to an object and a statement of the rights that a program has to manipulate the object. Programs may have capabilities recorded in special capability lists that represent their ability to reference objects external to themselves. Programs may pass capabilities to each other, but controls on whether a receiving program may in turn copy or pass a capability may be imposed. Although the addressing range of a program is constricted by small address spaces in instructions and by limited size of displacement or segment registers, a pointer is conceived as being very large, large enough to represent a systems-wide unique designation of the name of an object or segment.

An object, another program or a data object, is in the virtual memory of a program when a pointer to it is on its capability list. The presence of the pointer provides addressability and the statement of rights defines what a program can do to the object that may be addressed. Mechanisms in hardware translate object pointers to real memory locations by associating the segment name with a starting location. Mechanisms in the operating system may cooperate to determine what capabilities may be acquired by what programs.

The intent is to constrict the addressability of all programs to a limited private space, formalize the reference ability of external objects, and encourage modular programming. A proper way of thinking about a capability is as the name of an object. The capability concept is an extension of the segment concept described above.

## A Program

| | |
|---|---|
| Code Part | Contains coding whose addresses are limited to reference to code part or to private part. |
| Private Part | Contains scalar variables, working areas, and other spaces able to be referenced by addresses in code part. |
| Object Part (Capabilities) | Contains pointers to objects such as other programs, arrays, and record structures that are external to the program. A capability is a pointer to an object and a statement of the rights of this program in terms of ability to change, delete, execute, read, and so on. Also may contain rights to pass capabilities on to other programs. |

A private part of a program is uniquely established whenever the program is invoked so that various invocations of the program will have unique *incarnation records*. An incarnation record is basically an activation record, as described in Chapter 8. Capabilities need not be recorded in capability lists. Some systems permit the recording of a capability in the private part. There are various arguments, concerned primarily with performance and security, about whether capabilities should be organized into a separate list or recorded in the private part. The argument against object lists is performance due to a necessary additional level of addressing indirection. The argument against recording in private parts has to do with the possibility that certain implementations would allow for a capability to be permanently retained by a program, therefore enlarging its addressability and making reference to an object available to all subsequent callers.

The intent of such machine and operating system concepts is to organize architecture so that the structure of a machine more closely resembles the structure of well-formed modular programs in block structure languages. Some issues in how far an architecture should be extended to resemble the structure of programming languages are discussed in Chapter 12.

## QUESTIONS

1. What is the use of a page table?

2. What is demand paging?

3. What is a working set?

4. What use may an associative memory have in supporting virtual memory?

5. What role may virtual memory support play in protection?

6. What is a capability?

# Chapter 11

# Input/Output

## 1. INPUT/OUTPUT

The I/O architecture defines the visible interface a program uses to move data between a processor or memory and the I/O devices of the computer system. It must specify a method for designating a device to be used in an I/O operation. In addition it must describe the source of data to be moved for output or the target of data to be moved for input, the amount of data to be moved, and methods for determining when I/O operations have completed or encounter difficulties that keep them from completing.

The primary need for I/O is to provide a way for a computer system to exchange information with the human users of the system. The class of devices that output or accept as input human-usable forms is large and diverse. Among them are the following:

1. Printing devices to create hard copy. Printers vary in attributes, ranging from devices that are very much like typewriters to devices that can print thousands of lines of text per minute.

2. Ephemeral display devices that show text on a tube not unlike a television tube in general appearance. These displays may be formatted in various ways and may have the possibility of using multiple colors for highlighting.

3. Hard copy graphics devices such as line plotters that develop graphs and curves on paper.

4. Ephemeral graphics devices that can display pictorial representations of information using graphics techniques of some complexity. Some of these are capable of multicolored projections of three-dimensional images.

5. Multifunction printing and duplicating devices that can not only create text or charts but reproduce them in required amounts of copies.

6. Punched card devices that create card images that can be printed by printers not attached to a computer and that may have some top of card printing representing the data punched into the card.

7. Card readers capable of reading punched cards created on punched card machines and translating them into computer data representations.

8. Optical scanning devices that can read documents created by humans into a computer system. Sometimes these use specialized fonts or codes. Some devices undertake to read human handwriting.

132

The range of devices used by computers to communicate with human beings is expanding constantly, and the quality of interaction is improving as the population of users with different needs expands. In addition to the devices mentioned above, there is an anticipation that voice recognition and voice output will be part of the interaction repertoire within a foreseeable time period.

Concurrent with an increase in the ways computers can communicate with humans has been an increase in the tendency for interface devices to be *on-line* to a computer. That is, the computer talks directly to a user through the devices rather than preparing an intermediate form, such as a magnetic tape, that is then printed by a self-contained printing device for human use. The expansion of on-line interactive use of computers, as opposed to batch non-interactive use, has dramatically increased in the last decade.

Input/output is also used to permit a computer to communicate with other equipment, such as other computers, with telephone communications lines, and with various other industrial and scientific equipment. Thus a computer system uses I/O to receive signals from and to send control signals to an industrial robot or a rocket.

Another use of I/O is for a computer to communicate with devices that represent collections of data too large to keep in primary memory. The largest computer memories may currently contain up to 32 million bytes of data. Many on-line information systems, however, may keep multiple billions of bytes of data on storage devices accessable to one or more processor-memory complexes. On-line inquiry and data base systems rely on the use of *random access disk storage units*—also called *direct access storage devices* (*DASD*)—of differing speeds and capacity. A familiar application is the airlines reservation application where an airline represents flight legs and seat inventories on disk and reservations are made on-line by accessing this information as a result of activity at a display terminal.

The storage class devices include, in addition to disks and diskettes (smaller disk-like devices of limited storage capacity), rotating drums and magnetic tapes. These devices are addressed using I/O architecture and organization although they are not I/O devices in the pure sense of providing data exchange between a computer and a human or a computer and an instrument. They are intermediate storage units whose existence is due to certain economics of the technology of electronic storage and which are treated as I/O devices because of the details of the evolution of computing. They are probably most conveniently regarded as electronic filing cabinets. To respond to a request for data, a computer must interface with the human requesting data and the file in which the data is kept—much like a clerical employee who goes to a file to answer a question from another human being.

Actually, many architects and designers are considering future schemes where references to storage class devices would be made as if their storage space were an extension of the memory. An extension of virtual memory concepts would include all storage space in a *single-level store* where no application program ever issued direct commands for I/O but assumed that everything was directly addressable in the single address space of the single-level store. Mechanisms would exist in both hardware and software systems components to move data to and from various levels in a storage hierarchy in a manner invisible to the programmer who assumes that everything is in a virtual memory of unlimited size. These mechanisms would be extensions to mechanisms now used for virtual memory support. Current systems, with few exceptions, do not have this characteristic, and programmers do use I/O statements to reference data storage devices.

A fundamental and important characteristic of most I/O devices, be they human-oriented or intermediate storage units, is that they operate at considerably slower speeds than either processors or memories. Thus, references to data or interactions with humans consume enormous quantities of time as compared to computational functions within a processor-memory. For this reason the design of I/O is critical to minimizing delays and the design of I/O functions is frequently directly felt in the architecture. In this chapter we address a concept of architecture, in Chapter 20 we discuss design. The reader wishing a full view of I/O may go directly from this chapter to Chapter 20, which discusses aspects of interconnection, attachment, and data flow and provides some additional material on architecture.

## 2. DISK CHARACTERISTICS

Input/output architecture provides a definition of device characteristics as well as a specification of the forms and conventions used in a processor to accomplish input and output. Device architecture primarily addresses the appearance of data, both physical and logical, on I/O devices of various kinds.

Many of the features that characterize I/O devices in a system are design and implementation features that do not appear in the architecture. There are, however, a number of architectural aspects of I/O devices, for example, the conceptual grouping of data on a disk storage device or the definition of data addressing and storage and retrieval functions that can be applied to devices.

Perhaps the single most significant device architecture is that of disk storage units. A system may have a variety of disk storage units with diverse speeds, interconnection details, capacities, and physical structures. All of these will have in common the architectural concept of a rotating disk divided into *tracks* that are addressable as units from designated (or set of designated) record starting points. This concept is shown in Figure 3.

Each of the concentric circles in the figure represents a track of information. Each disk surface has an address, each track has an address. Systems conventions provide for the recognition of records on a track and may provide for the reading and writing of individual records recognized by a record header convention and the key of a particular record.

The distinguishing feature of disk storage devices is the ability to refer to data by key rather than by position in a sequence of records. The key or search argument is held in the system during a record search and compared against the

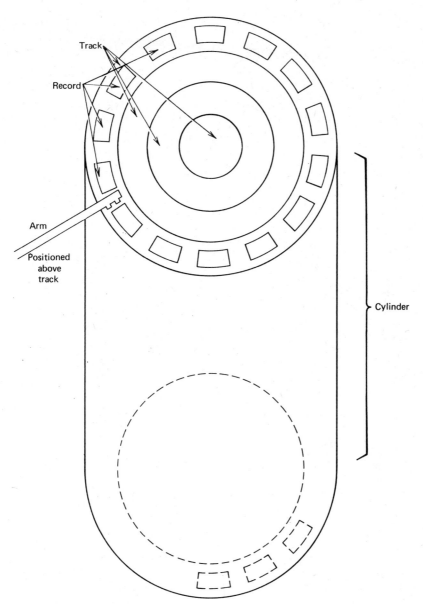

**Figure 3.**  Disk organization.

keys of records rotating underneath a read head. The time that it takes to read a record is the time required to move an arm to a particular track, *seek time*, and the time required for a track record header to rotate underneath the positioned arm, *rotational delay time*, followed by the time it takes to actually transfer data. The seek time may vary from 16 to 50 milliseconds for various disk storage units, the rotational delay time from 2 to 5 milliseconds.

Many disk storage units have the concept of a *cylinder*. A cylinder is a group of tracks on a number of surfaces that are all accessible at the same time by the same arm. This organization allows more data to be read at one time without positioning an arm.

Various disk units are built with special fast tracks that have more than one reading head associated with them. This cuts the rotational delay time because the distance that a record must travel to be found by a waiting read/write head is cut in half or in quarters, depending on the number of arms. There are also some head per track devices, which associate a head with each track in order to eliminate seek time.

## 3.  INPUT/OUTPUT ARCHITECTURAL FEATURES

A processing system must contain a set of instructions that are issued when it is desired to communicate with an I/O device of any type. These instructions provide:

1.  The I/O function that is to be performed. Typically READ or WRITE. READ transfers data from an I/O device to the processor-memory. WRITE transfers data from the processor-memory to a device.

2.  The designation of the device that is to be used for the I/O operation. A DASD type device organizes data into storage areas that are individually addressable. An I/O using this type of device must specify, in addition to the DASD unit, the address where desired data exist or are to be placed. Some devices do not have internal addressability of this type so that a reference to the device implies that the next data unit available is desired or the next sequential available space is to be used.

3.  The location in the processor-memory complex data are to be taken from (WRITE) or placed into (READ). It may also be necessary to indicate how much data is to be transferred between the processor-memory and the I/O device.

4.  Any synchronizing specification that may be appropriate. The I/O commands may specify whether or not an interrupt is desired and when the interrupt should occur. For example, an interrupt may be requested when the data transfer is complete.

5.  The recording format to be used, or the recording format to be expected from the device. Specification of binary, binary-coded decimal, and so on may be made.

Architectures vary as to how much of this information must be provided to constitute a complete I/O directive. The amount of information depends on the flexibility of the I/O functions. The more flexible the system, the more information is required. For example, on a computer system that can read or write only standard fixed size blocks of data it is not necessary to specify the amount of data to be moved. If a system only reads into a set of specialized fixed memory locations and writes from another it is not necessary to specify the memory locations. It may be necessary to move data to or from those special locations, but that is done with computer instructions used for memory location movement and not with I/O directives. If the system uses only one recording mode, if there is a fixed convention about interrupts, and so on, then the I/O specification need contain less and less information.

The simplest I/O architecture was on the first of the commercial computers, the UNIVAC. On that system it was possible to read or write on up to 10 tape drives and a computer console. Data transfer length was fixed to 60-word blocks for tape and to single words for the console. Any memory location that was a multiple of 60 could be specified. There was no interrupt, but an automatic buffering scheme with interlock to coordinate I/O. To read a 60-word block into memory locations starting at 880, from tape unit 2, one issued the single instruction 52 880.

The architectures of later machines allow one to specify a great deal more than that and are certainly more flexible and potentially more efficient. The price payed for this, however, is an enormous amount of software support to enable a program to deal with the complexity of flexibility.

### 4. BASIC PROCESSOR I/O ARCHITECTURE

A somewhat simplified model of the 370 generic I/O architecture is used as an example of I/O architecture. It is a rich architecture that enables us to look at many facets of I/O. Chapter 20 describes some other architectural approaches in connection with some concepts of design.

The 370 architecture is characterized by flexibility. As a consequence, a great deal of information must be provided to form a complete I/O directive. This information is provided in a number of places. Of primary importance are the following:

> Start IO instruction (SIO)
> Channel address word (CAW)
> Channel command word (CCW)

Together these provide all of the information required to initiate an I/O operation. The SIO designates the I/O device involved in the operation. The designation contains two components: the designation of the device and the designation of the path to the device. The concept of the path to the device introduces the notion of connection methodology. Figure 4 shows a population of devices that are attached to *control units* which, in turn, are attached to *channels*. The full designation of a device includes a designation of the channel and control unit as well as the device. Thus the SIO specifies how to get to the

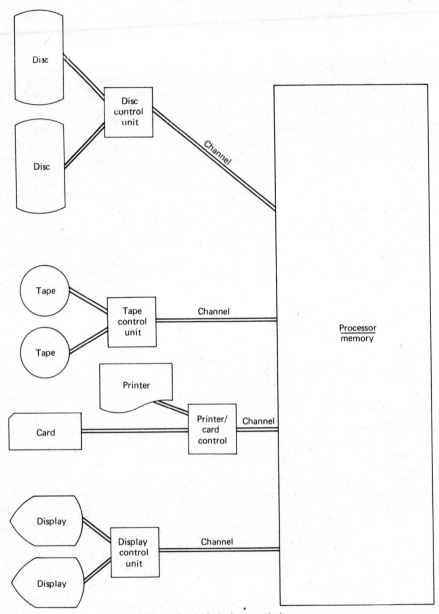

**Figure 4.** I/O device population.

device as well as the device itself. The need for this derives from the possibility, in large configurations, that a device may be accessible from more than one channel or more than one control unit. The alternative paths are provided for reasons of performance or reliability.

The execution, in 370 architecture, of an SIO causes the system to refer to a fixed location in memory that holds a channel address word (CAW). The CAW serves two purposes. It contains a protection key (see Chapter 5) for a program undertaking I/O. This protection key is matched against the protection key of the memory locations that hold or receive data for the I/O. If the keys do not match then the I/O operation will not be performed.

The CAW also contains the address of a list of channel command words (CCW) that direct the details of the I/O operation. A CCW word contains:

1. The exact I/O operation to be performed, for example, READ or WRITE.
2. Subfunctions related to the device it is to be performed on. For example, setting tape writing densities in terms of bits to be written per inch or specifying paper advance for a printer.
3. The memory locations and the amount of data that is to be transferred between the I/O device and the processor-memory.

A simple I/O operation may be defined with a single SIO, CAW, and CCW. It is possible to define more complex I/O operations by associating a list of CCWs with an SIO and a CAW. For example, a single I/O operation may specify the reading of 1000 bytes into locations 9000 to 9999, followed by the reading of 500 bytes into locations 15000 to 15499. This read of 1500 bytes into discontinuous memory locations would be accomplished by the use of two CCWs, one for the 1000 byte read, another for the 500 byte read. This kind of chaining is called *data chaining*. Data chaining indicates that, while various memory locations are to be used in accordance with CCW addresses, the function that is to be performed will be the function specified in the first CCW of the chain.

CCWs may also be chained in such a way that the commands (READ, WRITE, etc.) may change from CCW to CCW across a single conceptual I/O operation under the control of a single CAW and a single SIO. This form of chaining is called *command chaining*.

Thus very rich and complex I/O operations can be constructed by the use of CCW chains. As long as the device reference is the same and protection keys do not change, composite reads, writes and other functions can be performed involving discontiguous areas of memory space.

In versions of the architecture that include virtual memory, special provision is made for I/O through the use of *channel indirect data*. This is an application of indirect addressing to I/O. With this feature the address field of a CCW indicates the address of a supporting software table that partitions I/O spaces into page frames. Thus each CCW address field is itself supported by a list of page frame addresses. The operating system must pin these pages in memory, that is, not allow them to be paged out, during an I/O operation.

There are variations on the I/O architectures from architecture to architecture, naturally. The flexibility of defining blocks of different size and moving into discontinuous areas and the visibility of intermediate units (channels and control units) along the path to the device all change as a function of the price level of a system and the complexity and richness of its I/O capability. There is an intense interplay between architectural and design concepts in the I/O area.

### 5. OPERATING SYSTEMS SERVICES

The issuance of an I/O instruction in a contemporary computer system of large or intermediate size is the end of a rather rich set of events involving various aspects of the operating system, the programming languages, and the hardware.

In many architectures the issuance of an I/O instruction is limited to control state so that an application program must call on the operating system to issue I/O directives on its behalf. Reference to control locations in lower memory that are associated with I/O is also limited to control state so that response to a completed I/O may also necessarily involve the operating system.

In addition to issuing I/O commands and responding to I/O completions, the operating system plays other important roles in the I/O flow. The operating system provides two major services:

1. The operating system, or extensions to the operating system called *access methods*, provide simpler interfaces for an application. Thus an application can issue software macro commands like READ, WRITE, GET, and PUT that are translated by the operating system into hardware architecture forms. For the application program the effective interface to I/O is defined by its image of the forms provided by the operating system and not by the hardware architecture. The parts of an operating system involved in this are the access methods and the *I/O supervisor*.

2. The operating system provides information about the location of referenced files as part of its function of allocating devices and device space to data files. In a contemporary system, a program need not know exactly where the files it wishes to reference are located. Any file will either already exist on the system or be created by the program. Files already existing are noted in a catalogue or directory maintained by the operating system. Each such file is represented by an entry that provides the file name and its location in the device population. These entries are kept on a disk.

   If a program wishes to use a file on a system where preexistent files are not permanently mounted, the operating system will determine the device onto which the file should be mounted and issue a directive to operators to mount the file on the device. When a program creates a new file, the operating system will allocate permanent or temporary device space for it.

   The assignment of devices to files is performed by a part of the operating system called the *allocator*.

When a program is started, the allocation of files that it references must be made known to the I/O Supervisor in some convenient way. The operating system constructs *device assignment* tables in memory that associate a file name with a file device allocation. This table may be constructed when the program starts or when the program issues an OPEN command to the operating system naming a file that it is to use. After an OPEN, a program may GET, PUT, READ, or WRITE a file.

The program as written may specify the symbolic name of a referenced file. The compiler will treat this name in such a way that when reference to the file is made the table representing the file and its allocation is referenced. The table may contain specific CAW and CCW type words that have been generated by the operating system as a result of the allocation of the file and memory space as well as the compiler's interpretation of the record structure involved in the I/O operation. This function of an operating system provides *device independence*. Device independence permits a program to be written and compiled without knowing the specific physical devices where referenced files may be accessed. It provides considerable flexibility in shared system use as well as flexibility for running an individual program.

| | |
|---|---|
| PROGRAM STATEMENT: | WRITE File A |
| COMPILER OUTPUT (Symbolic): | SUPVCALL (Write, A) |
| DEVICE ASSIGNMENT TABLE: | A:  SIO: Device 191 |
| (Symbolic) | CAW: Key, CCW List |
| | CCW: Write 100000, 1000, * |

The above is a simplification and generalization that does not represent the forms used by any particular operating system running on 370 architecture. We are showing only the general nature of the structures and forms that are involved. The actual nature of tables and structures used by an operating system to reflect device assignment may be considerably more complex and represent more information than what we have shown. In the above table "CCW List" is the address of the beginning of the CCW list and "*" indicates that only one CCW is used.

The SUPVCALL instruction specifies that it wishes to execute a write onto File A. At the time the program was allocated space, a memory assignment was made for A that provided space for the device assignment table that represents File A. The I/O supervisor accesses this table and finds in it the SIO, CAW, and CCW necessary to perform an I/O. These were formed when File A was allocated to device 191. At the same time the program was allocated memory space so that data written to File A would come from memory location beginning at program relative 100000. The length of the write was determined, possibly, by a DECLARE statement, by specification in the WRITE, or by some associated formatting statement in the programming language.

The example above showed a direct link between a program and the I/O supervisor provided by the compiler. Actually the link generated by the compiler may be to an access method and a call to the I/O supervisor issued only when the access method finds it necessary.

An access method provides two fundamental services, buffer management and record advance. It is common for a program to designate that a number of areas in memory are to be used as places to put data or to take data from. On a model of an architecture that permits I/O to proceed in parallel with computing, it is usual for the program to be working in one area while another area is being filled or drained by the I/O device. The determination of which of these buffer areas a particular I/O request is to use is made by the access method.

When the I/O operation specified by a program is READ or WRITE, the access method always calls the I/O supervisor because a real physical I/O operation is requested. A program may issue GETs and PUTs associated with record advance. Record advance assumes that records have been *blocked*, which means that any physical read or write will move a unit of data that contains a number of individual records. Processing of input is accomplished by a program issuing a GET; the access method provides the address of the next record in the block. The processing of output involves the program issuing a PUT that causes the access method to add a record to a block that is being prepared for output. The blocks of records may be buffered. The access method will issue a request for physical input or output only when a block of input is depleted of records and when a block of output is completed. At such times the access method will call on the I/O Supervisor to undertake the I/O operation.

## 6. INPUT/OUTPUT SUPERVISOR FLOW

In this section we look more closely at the nature of an I/O supervisor. We describe the general characteristics of an I/O supervisor that might run on 370 architecture. The description, however, is not an exact description of the I/O supervisor of any of the actual IBM operating systems for 370 architecture.

The primary function of an I/O supervisor is to submit I/O requests and to respond to interrupts that indicate that I/O operations have been completed. When a problem program or an access method calls on the operating system for an I/O it is common to suspend the program that has requested the I/O. This is done on the assumption that the program cannot proceed until the I/O request is satisfied. The suspension involves the marking of the process control block as non-dispatchable and the saving of the status of the program. A process control block is a software defined structure that represents the status of a running application program.

The I/O supervisor places the appropriate CAW in a fixed location in memory and issues an SIO. If the device is busy, or the path to the device is not accessible, an interrupt will occur. The I/O supervisor then places the request for I/O on a queue for the device and enters the dispatcher for the execution of another problem program.

Over time an interrupt from the I/O hardware indicates the completion of the I/O that has delayed a queued I/O request. When an I/O completion interrupt occurs, the first-level interrupt handler (see Chapter 9) calls on the I/O supervisor. The I/O supervisor has two functions to perform:

1. It must post completion of the I/O. This involves determining what program caused the initiation of the now completed I/O and making some indication in its process control block that the I/O is now complete.
2. It inspects queues of I/O requests for an I/O request that wished to use the now free device. When it finds a request it issues an I/O by moving the relevant CAW word to its proper place in memory and issuing an SIO.

When response to I/O completion and issuance of new I/O is performed, the I/O supervisor will either return to an application program or to a dispatcher that will select an application program to run next on the processor. The conditions under which the I/O supervisor returns directly to a suspended application program or to the dispatcher are defined by the *preempt resume* and *scheduling* strategies of a particular operating system.

Operating system and I/O design and organization will determine the details of these functions. Variations include the number and organization of queues, the exact nature of the representation of an I/O request, and the algorithms used for submission of I/O in either first come, first served or some priority order. Despite the variations, however, the flow described is quite typical. The entire I/O flow comes from a collaboration between operating system and compiler superimposed on the basic processor I/O architecture. The programming language specifies I/O data structures symbolically, the compiler creates linkages to an access method, and the access method links to the I/O supervisor for performance of I/O. The operating system also determines the allocation of files to devices and records these allocations in various tables conveniently used for device reference. In addition some I/O directive generation may be performed by the operating system to reflect allocation decisions.

We mentioned at the very beginning of the chapter that there may be simpler environments due to a reduction in flexibility. Different architectures may fix the block size and eliminate certain options and addressing requirements (such as the address of the channel to the control unit) so that the basic I/O instruction may be reduced to a single word. Similarly, operating design may be simplified to reduce the number of options and the operating system flow. The operating systems designers may choose not to make available all of the options that may potentially exist in the I/O architecture.

An essential point to remember about I/O architecture for many systems is that the visible features of the I/O function are defined primarily as software interfaces that hide the hardware features of the architecture. These software features include concepts of allocation and service levels that are operating system translated into the I/O hardware architecture of the machine.

## 7. CONTROL LANGUAGES

We have so far discussed I/O as a composite architecture involving operating system, hardware, and compiler. There is, however, one additional aspect that should be discussed. Most operating systems have some kind of *control language* that is used to request the running of a program and to describe the resources that a program will need to run.

The modern programmer, when facing I/O, confronts the intersection of the programming language he is using and the command or control language associated with the operating system of the machine. Different programming and control languages have different, sometimes overlapping, abilities to describe I/O environments. It is usual for the programming language to provide descriptions, via DECLAREs, of the record structures and formats to be associated with a READ or WRITE. The control language provides information about the allocation of the data on devices and the nature of the devices on which allocation of files is to occur. Thus, the definition of the characteristics of a file, of the disposition of a file after the program ends, and so on, must be made in the command language. While some operating systems attempt to provide simplifications of these control definitions or establish limiting conventions that make them unnecessary, it is still a factor in a complete understanding of I/O flow.

Consider the following example:

```
RUN PROGRAM A
FILE A, 50000 Records, New, Permanent, Disk, 100-Byte Record.
```

This control language statement tells the operating system that Program A is going to create a new file of 50,000 records that is to be permanently assigned space on a disk associated with the system. The specific disk is left up to the operating system. The statement also includes information that each record of the file will be 100 bytes long.

This statement gives the operating system enough information to allocate space for File A and enter its existence into a system catalogue. As we have seen, when Program A was compiled it provided some information to the compiler about the record structures used in a File A record. It also may issue GETs, PUTs, READs, or WRITEs to File A. The compiler generated linkage to an access method and provided access method coding to be included in a load of Program A. The compiler also included a command to the access method to OPEN File A. There are many variations in the details of what happens when at the time Program A will run.

One typical sequence is as follows. At the time the operating system sees the FILE statement in the control language it does allocation and enters the allocation in the file catalogue. When the program issues an OPEN statement (compiled as a call to the operating system), the operating system develops the device assignment tables, generating at that time the instructions that will actually be used to perform the I/O.

One interesting problem is that the intersection of programming languages and control languages is not standard. Different programming languages allow different kinds of data declaration, some subsuming anything that might be done in a control language, some obviously depending on an implied control language for file definition. Thus a programmer faces different requirements for I/O definition across programming and control languages depending on the language and the system he is using.

## 8. HIGHER LEVELS OF INPUT/OUTPUT SUPPORT

This chapter has undertaken to introduce the basic nature of I/O on an architecture of some flexibility and to discuss the role of software in support of I/O. Chapter 20 completes the discussion, adding more examples of architecture and providing an introduction to some design and organization concepts. The discussion has made simplifications at many points, particularly in the description of operating systems functions. However, a proper introduction to the nature of I/O has been established and one hopes that an area that seems most mysterious and threatening has been rendered less intimidating.

One area that the discussion has not touched on at all is the nature of software programs that are generically called *data base management systems*. These systems typically provide considerably more support and function at the application–program interface. In place of communicating with an access method or an I/O supervisor, an application program communicates with a data base manager. The data base manager provides a view of data and data relationships considerably more rich and abstract than that provided by the operating system. The organization of units of data, the fields and values that can be retrieved by a request for data, are completely divorced from the physical representation of data in the system. Diverse and multiple logical data relationships can be defined by a program that the data base manager maps into physical requests to be processed by the operating system.

## QUESTIONS

1. What kind of information must be provided to do an I/O?

2. What is DASD?

3. What is rotational delay time?

4. What information is provided in a S/370 CCW?

5. What is device independence?

6. What are major functions of an I/O supervisor?

## BIBLIOGRAPHY

This bibliography provides source material for Chapters 1–11.
*Communications of the Association for Computing Machinery* is shown as *CACM*.

Bell, C. G., and A. Newell. *Computer Structures: Readings and Examples*, McGraw-Hill, New York, 1971.

Borgerson, B. R., M. L. Hanson, and P. A. Hartley. The Evolution of the Sperry Univac 1100 Series: A History, Analysis and Projection, *CACM*, Vol. 21, No. 1, January 1978, p. 25.

Case, R. P., and A. Padegs. Architecture of the IBM System/370, *CACM*, Vol. 21, No. 1, January 1978, p. 73.

Chu, Y. High-Level Computer Architecture, *COMPUTER*, Vol. 14, No. 7, July 1981, pp. 7–9.

Cragon, H. G. The Elements of Single-Chip Microcomputer Architecture, *IEEE Computer*, October 1980, p. 27.

Denning, P. J. Third Generation Computer Systems, *ACM Computing Surveys*, Vol. 3, No. 4, December 1971, p. 175.

Data General Corporation, How to Use the Nova Computers, DG-NM 5, April 1971.

Donovan, J. J. *Systems Programming*, McGraw-Hill, New York, 1972.

Foster, C. C. *Computer Architecture*, Van Nostrand Reinhold, New York, 1970.

Foster, C. C. A View of Computer Architecture, *CACM*, Vol. 15, No. 7, July 1972, p. 557.

IBM Corporation. IBM System/370 Principles of Operation, *GA22-7000*, 7th ed., March 1980.

INTEL Corporation. 8080 Assembly Language Programming Manual.

Levy, H. M., and R. E. Eckhouse, Jr. *Computer Programming and Architecture, The VAX-11*, Digital Press, Maynard, MA, 1980.

Lorin, H. *Parallelism in Hardware and Software: Real and Apparent Concurrency*, Prentice-Hall, Englewood Cliffs, NJ, 1972.

Lorin, H., and H. Dietel. *Operating Systems*, Addison-Wesley, Reading, MA, 1980.

Myers, G. J. *Advances in Computer Architecture*, John Wiley & Sons, New York, 1978.

Morse, S. P., B. W. Ravenel, *et al.*, INTEL Microprocessors—8008 to 8086, *IEEE Computer*, October 1980, p. 42.

Organick, E. I. *Computer System Organization—The B5700/B6700 Series*, Academic Press, New York, 1973.

Presser, L., and J. R. White. Linkers and Loaders, *ACM Computing Surveys*, Vol. 4, No. 3, September 1972, p. 149.

Rosin, R. F. Supervisory and Monitor Systems, *ACM Computing Surveys*, Vol. 1, No. 1, March 1969, p. 37.

Russell, R. M. The CRAY-1 Computer System, *CACM*, Vol. 21, No. 1, January 1978, p. 63.

Schoeffler, J. D. IBM Series/1, The Small Computer Concept, IBM Corporation General Systems Division, Atlanta, GA, 1978.

Stone, H. S. Introduction to Computer Architecture, Science Research Associates, 1975.

Wulf, W. A. Compilers and Computer Architecture, *COMPUTER*, Vol. 14, No. 7, July 1981, pp. 41–48.

# Chapter 12

# The Power of
# an Architecture

## 1. CONCEPT OF ARCHITECTURAL POWER

The previous chapters have described the concepts that are part of computer architecture and have provided examples from specific architectures to illustrate various concepts. There are a large number of different computer architectures. Questions naturally occur about whether there is a "best" computer architecture and what kind of measurements of architectural efficiency might be used to compare one architecture to another.

We must be careful to distinguish the question of best architecture and the issues of comparative architectural efficiency from issues of fastest computer. The raw speed of a computer is a function of its design and implementation as well as its architecture, and we need some measures other than raw speed, or even price/performance, to measure one architecture against another.

It is certainly true that architecture, design, and implementation influence each other. Different architectures require design and implementation techniques of differing complexity and expense to achieve various speeds. The issue of architectural efficiency is sometimes argued in terms of the complexity of design implied to achieve certain performance levels or instruction execution rates. One measure of an architecture seems to be the efficiency with which it can be represented in the creation of a real computer and the number of circuits required to achieve high-performance models of the architecture.

Another approach to assessing architectural efficiency is the ease with which a compiler can be developed that can generate "good" coding for the architecture. Thus some contend that symmetry (see Chapter 6) is important because it reduces the burden of assessing special cases and selecting coding patterns to fit special cases of data length and representation.

But setting aside problems of physical realization and compiler design, there is considerable interest in measures of architectural efficiency assuming the best possible compiler and no organization or implementation constraints. Is there such a thing as an ideal architecture? Is there a way of defining a method of recognizing it? Immediately one encounters the problems of clearly distinguishing between architectural and design concepts, of constructing categories of classification for the attributes of an architecture, and of developing metrics for measuring architectural efficiency. These problems are confounded by the reality that computers are presented with work that itself may have very different characteristics so that an architecture that is better than others for one type of problem may not be as good for another. For example, an architecture that allows very efficient coding of certain mathematical algorithms may be weak in moving rather complex data structures between memory locations. A program that undertakes to find the sine of a simple single-word value may look very efficient, but a program that undertakes to sort variably sized records may look rather inefficient in the architecture.

Almost all of the argument and proposed methodology for comparing architectures to each other contains some recognition of this variation in work load type or involves heated argument about what kind of work is most "typical." For example, there is some disagreement about what kind of arithmetic expressions are most commonly found. The most efficient architecture for evaluating $A=B$ does not seem to be the most efficient architecture for evaluating $A=B+C*D/E$. Consequently, arguments develop about the distribution of the occurrence of assignment statements as regards the number of operators and variables in a statement.

This chapter undertakes to introduce various approaches and arguments related to measurements of architectural efficiency. We consider two dimensions of this problem. First we describe various searches for important architectural attributes and metrics, demonstrating some of the arguments that accrue to determining the best architectural concepts. Second we describe some of the issues associated with current thinking in the area of the *level* of an architecture. By level one means the extent to which the architecture of a computer should be moved in the direction of assuming functions that have been traditionally performed by an operating system or a compiler. There are various opinions about the desirability of raising the *hardware/software interface* so that less operating system function is represented in conventional programming and the structure of the computer more closely resembles the concepts that are presented in higher-level programming languages.

## 2. THE ARMY–NAVY STUDY

Processors are sometimes characterized in terms of instructions executed per second. This rate may be called the MIPS rate (for processors that execute millions of instructions per second) or the KIPS rate (for processors that execute thousands of instructions per second). High-speed processors designed for scientific computing are sometimes characterized by the FLOPS rate (floating point operations per second).

Clearly the instruction execution rates of processors with different instruction sets, addressing forms, and register populations do not form a basis for architectural comparison. There are techniques that try to normalize the power of architectures so that machines of different architecture can be compared using implementation independent metrics. There have been attempts to derive standard measures of architectural power so that comparative statements between machines of different architecture can be made.

One of the most widely known attempts to create a framework for the evaluation of architectures was undertaken by a joint U.S. Army–Navy Committee in the mid-1970s. The goal of the effort was the establishment of a process by which an architecture could be selected for use in military computers. The motivation for the selection of an architecture was the ability to make maximum use of existing software, to avoid the proliferation of new architectures, and to use architectures whose attributes were well known. The intent was to qualify architectures that could lead to the use of a standard computer family of diverse performance characteristics.

The criteria developed for architectural comparisons were applied to a set of commercial and military computers by the joint committee and have been independently applied by others to architecture not in the original study. There is by no means agreement that the criteria are well understood and easily applied, but the effort is interesting because it is perhaps the most extensive attempt to construct architectural measurements.

### 2.1.  Absolute Criteria

The criteria developed for an architecture are of two basic types, absolute and quantitative, in the terminology of this effort. Absolute criteria include:

1. Virtual memory support.
2. Protection.
3. Floating point support.
4. Interrupts and traps.
5. Subsetability.
6. Multiprocessor support.
7. Control of I/O.
8. Extendability of instruction set.
9. Read-only memory.

The intent of the absolute criteria was to qualify an architecture in terms of its general function and completeness. The requirements for an architecture, in terms of the absolute criteria, are as follows (the characterizations are mine, not the committee's):

1. *Virtual Memory Support.*   There must be an ability to translate virtual memory addresses into real memory addresses in the spirit of Chapter 10. Some kind of paging and/or segmentation mechanism must exist.

2. *Protection.*   There must be a way to protect specified areas of memory and an ability to ensure that a program may not gain unauthorized access to protected memory areas or to privileged operations by increasing its addressing or execution rights illegitimately.

3. *Floating Point Support.*   The architecture must provide instructions for floating point operations.

4. *Interrupts and Traps.*   It must be possible to guarantee correct return to an interrupted program under any condition of interrupt or sequence of interrupt classes or types.

5. *Subsetability.*   The architecture must be able to be partitioned into defined subsets so that it is operable without virtual memory support, floating point instructions, decimal instructions, or the protection mechanism.

6. *Multiprocessor Support.*   The architecture must contain instructions that will enable multiprocessor configurations to be used. Such instructions are typically those that provide for the synchronization of more than one processor usage of shared data areas or shared code.

7. *Control of Input/Output.*   The architecture must contain all the features necessary for the complete control of I/O devices.

8. *Extendability of Instruction Set.*   There must be some clear possibility for adding instructions to the instruction set.

9. *Read-Only Memory.*   The architecture must be able to execute code from memory locations that cannot be modified.

There is some difficulty in precise interpretation of all of these criteria and some question about whether all of them are significant architectural features. Some architects, for example, would argue that extendability of an instruction set, adding additional instructions and thereby increasing the complexity of the architecture is not a good idea at all, that, to the contrary, the fewer instructions in an architecture the better. Similarly, the distinction between read-only memory and protection is not clear, as it is not clear that read-only memory is an architectural, as opposed to an organizational, feature. Others have found it impossible to make sense of the requirement to control I/O, since it seems inconceivable that an architecture would not contain enough logic to run I/O devices in a meaningful way.

These comments raise some issues of how clearly distinctions between architectural and organizational features have been made. They are not intended to do more than make the reader cautious. The intent of the study was to isolate purely architectural issues, and any possible failures are as interesting as successes since they suggest the difficulty of the undertaking. The absolute criteria are apparently sufficiently discriminating to have eliminated six of nine candidate architectures before the application of additional criteria.

These absolute criteria were combined with a study of the availability and maturity of software support for the architecture and the development of a method for weighting various factors in the evaluation of an architecture. Clearly the absolute criteria, the maturity of software support, and the evaluation method are not related only to the notion of architectural power but also to the notion of architectural desirability in terms of future growth, programming expenses, installation expenses, and so on.

## 2.2. Quantitative Criteria

Architectures that passed the qualitative absolute criteria were subjected to a series of quantitative measures that are more directly aimed at making comparative statements about architectural efficiency and power. Among the quantitative tests are the following:

1.  Size and organization of virtual address space.
2.  Size and organization of physical address space.
3.  Fraction of instruction space not used.
4.  Size of processor state.
5.  Usage base.
6.  Direct addressability.
7.  Virtualizability.
8.  I/O initiation.
9.  Interrupt latency.
10.  Subroutine linkage.

The meanings of these elements of measure are as follows:

1. *Size of Virtual Address Space.*    The number of bits in the virtual address space is one measure. An additional measure is the number of addressable units in the virtual address space. Thus a virtual memory that had $2^{32}$ bits and was addressable in bytes would have $2^{28}$ bytes. Goodness here is clearly large address space with very granular addressing. A $2^{32}$-bit memory addressable only as 32-bit words would be less desirable than addressability to bytes.

2. *Physical Address Space.*    As above, but related to the largest configuration of memory available with the architecture rather than the largest address it is possible to develop on the architecture. Large physical address spaces are good because diminishing ratios of virtual to physical address space imply better computer performance due to lower levels of paging activity. Whether this is a purely architectural feature is not clear since one may argue that the architectural decision involves addressability and that physical memory size is an organizational issue.

3. *Unassigned Instruction Space.* This is a quantification of the absolute measure relating to extendability. The measure derives from an algorithm that projects the potential number of new instructions from the pattern of used and unused bits in an instruction word. It will be influenced by the way instructions are coded, the use of secondary operation fields, the number of instruction formats, and so on. It is considered to be an important indication of the growth potential for the efficiency of the architecture.

4. *Size of Processor State.* This is a measure of the efficiency with which task switching operations or interrupt handling can be performed. The measure addresses the number of bits that must be represented in a processor to reflect the data that must be available to a program if it is to run, be suspended, and then be resumed. The measure has two basic components, the number of bits required to fully represent the status of a running program and the number of bits that must be moved from the processor to memory to suspend and then resume a program. This metric of the number of bits that must be moved between a processor and memory to support a certain function is an important concept in measuring architectural efficiency. We will encounter this measure repeatedly in various guises. In various forms it is almost the central concept that links various approaches to architectural efficiency. In general an architecture that moves fewer bits between processor and memory to perform certain functions is considered more powerful than an architecture that moves more bits.

5. *Usage Base.* This represents the number and dollar value of the delivered units of the architecture. It is perhaps an attempt to invent some surrogate number to indicate software maturity and architectural stability. Clearly the measure has little use in measuring architectural efficiency.

6. *Direct Addressability.* This is the largest number of bits that can be transferred between processor and memory by use of a single instruction. The effect on architectural efficiency is the minimization of the number of instructions required to move fields of various lengths. An architecture that can move 32 bits with a single instruction is considered more powerful than an architecture that can move only 8 bits with a single instruction.

7. *Virtualizability.* The virtualizability of an architecture lies in the ability to define virtual machine operating systems by building software that can use defined subsets of the privileged instruction set.

The concept of a virtual machine is an extension of the concept of virtual memory. The virtual machine concept involves the ability to define disk storage spaces, consoles, telecommunications lines, printers, and so on in such a way that a defined set of resources may be put under the control of a particular operating system. Thus a user is provided not only with a virtual memory space but with virtual disks, and so on, running in an environment that gives the user the impression of a dedicated stand alone machine. One use of the concept is to permit multiple different operating systems to share the same machine.

The concept is supported, characteristically, by the creation of an underlying monitor program that provides interfaces to the operating system. Various hardware architectures have different characteristics that determine the feasibility of defining virtual machines. The nature of the protection mechanisms, the details of virtual memory support, interrupt systems, and even some features of the instruction set will affect the feasibility of defining virtual machines on a real machine base.

8. *Input/Output Initiation.* The number of bits that pass between the processor and memory in order to output 1 byte to an I/O device.

9. *Interrupt Latency.* The number of bits that must be transferred between memory and a processor of the system in the interval between the receipt of an interrupt and the processing of the interrupt. This measure is distinct from the processor state criteria in that it does not include the storing and restoring of program status information. The bits moved include those structures that must be automatically stored in memory on the occurrence of an interrupt. Examples would be the PSW storage into memory, the movement of a new PSW to an instruction counter, the storing of any address control registers, and the fetching of new address control values. The factor seems to be considered good in the context of the Army-Navy study if the bit count is low. That implies a minimum period of delay before interrupts can be handled.

10. *Subroutine Linkage.* The number of bits that must be transferred between the processor and memory to support a parameterless call of a subroutine and the return to the caller.

These quantitative criteria are analyzed for an architecture, and they have criteria weights applied to them. The criteria weights show the relative importance to the selection process that each criterion is to have. The evaluation process consists of the application of the absolute criteria and the application of the weighted quantitative criteria to the architectures that pass the absolute tests. Certainly the weighting of importance is an area of potential argument and divergent opinion. The relative importance of the quantitative criteria used in the Army-Navy study is as follows:

1. I/O initiation.
2. Direct instruction addressability.
3. Interrupt handling bit movement.
4. Subroutine linkage.
5. Physical address size.
6. Unassigned instruction space.
7. Processor state space.
8. Virtualizability.
9. Addressable physical units.

10. Addressable virtual units.
11. Subroutine linkage in a subset architecture.
12. Processor state space in a subset architecture.
13. State space movement.
14. Size of virtual memory space.
15. State space movement in a subset architecture.
16. Number of computers delivered.
17. Dollar value of computers delivered.

This ranking of importance may give some indication of the attributes of an architecture that are interesting in evaluating power. Caution is necessary because the list represents a composite of the Army list and the Navy list, which are not identical in either weightings or order of importance.

## 3. MEASURES OF AN ARCHITECTURE

The Army–Navy study introduces us to the idea that a fundamental measure of a computer architecture is the number of bits that must be moved from processor to memory in support of certain activities. The bit count is a function of the number of instructions that must be executed to do an activity, the size of the instructions, and the number of data bits that must be moved.

Some abstract "most powerful architecture" will execute a minimum number of instructions, each of minimum size, and transfer only those data bits that are required to perform the activity. By inference, powerful instructions reduce the number of instructions, very effective addressing formats reduce the size of the instructions required, and very efficient data representation and addressing reduce the number of data bits that must be transferred. The more powerful architecture will also pass another figure of merit for an architecture, the amount of memory space required to represent a given program. An architecture that provides for the fewest, shortest instructions will have greater *code compaction* than an architecture that requires more and longer instructions. Since memory is an important component of computer cost it is better to represent programs in the smallest possible space.

Two basic measures of architectural efficiency, therefore, are the number of bytes required to represent a program and the number of bits moved between processor and memory in the execution of a program. To these there is sometimes added the number of bits that must move between the registers of a processor during the execution of the program. This last measure accounts for the necessity of moving values from one register to another in the execution of various algorithms, either because the architecture has special-purpose registers that require movement, for example, from an arithmetic register to an index register, or in order to avoid making memory references.

We see that the best architecture allows for the representation of programs in minimum space and minimizes the number of bits that are moved between processor and memory and between registers of the processor. This concept of an ideal architecture provides a concept of space and work required to perform the functions of computing.

It might at first appear that the problem is solved, but, unfortunately, because of the diversity of work that is presented to a computer, an architecture that minimizes space and work for one class of problem may not minimize it for another. It is for this reason that we find divergences in architectural style depending on the architect's concepts of the primary use of a machine based on the architecture. An architecture aimed at high-speed floating point computation will diverge from an architecture aimed primarily at handling simple transactions on complex data structures primarily in coded decimal. Although both architectures may do either job, the space and work figures of merit may reverse themselves across different patterns of programming and problems.


## 4.   FUNCTIONAL INSTRUCTIONS

Some work in measuring the quality of an architecture investigates the notion of the number of instructions that do not really do anything except set up for the execution of significant instructions or control the sequence of the machine. A significant or *functional* instruction is defined to be an instruction that really accomplishes a change in data associated with the running of the program. Thus an ADD, SUBTRACT, MULTIPLY, or DIVIDE instruction performed on problem data is a functional instruction.

Many instructions executed by a program are used only to move data to a place where functions can be performed or where the results of functions can be recorded. Thus LOAD and STORE instructions, MOVE instructions whose only purpose is to move data to buffer or working areas, and so on are not useful instructions, and a superior architecture requires fewer of these instructions than an inferior architecture.

Similarly, instructions like COMPARE and JUMP are procedural instructions whose use does not contribute to the transformation of data. A superior architecture minimizes the use of procedural instructions. Of course a difficulty is that instructions like COMPARE and JUMP may be functional in some programs and not in others. In a sorting program they are at the heart of the concept of the algorithm and are certainly functional for that program. In a loop construct they would be procedural in the sense that they are used only to control program flow and might be eliminated or reduced in a machine with vector instructions.

Despite some uncertainty about the absolute functional or nonfunctional classification of instructions, a proposed figure of merit for an architecture is the ratio of nonfunctional (data movement and procedural) instructions to functional instructions. The difficulty in using the measure is that across the population of potential programs nonfunctional instructions may apparently be functionally significant in some programs.

The most obvious effect of an architecture that reduces the number of nonfunctional instructions is to reduce the number of instructions actually executed. By additional careful design an architecture that reduces the number of instructions may also reduce the program size.

$$
\begin{aligned}
&\text{LOAD R1, A} \\
&\text{LOAD R2, B} \\
&\text{ADD R1, R2, R3} \\
&\text{STORE R3, C}
\end{aligned}
$$

By measures of functionality, this code contains only one functional and three nonfunctional instructions.

Consider the following instruction:

$$
\text{ADD A, B, C}
$$

This is clearly superior by the measure of ratios of functional to nonfunctional instructions. To measure whether the single instruction is superior by the space required and the bits moved we must postulate some further environmental features. The reader may remember that we first undertook this kind of discussion in Chapter 5 where we discussed conditions under which various addressing forms were going to be more efficient under various assumptions of address space requirements and register populations. At that time we determined that counterexamples could be found for the efficiency of various architectures as a function of the perception of normal work flows.

The point was that a clearly stated figure of architectural merit, the number of functional vs. nonfunctional instructions, is not only unstable because of workload assumptions but may lead to conclusions about architectural merit inconsistent with other measures of architectural efficiency. That is, a bits moved metric and a functional/nonfunctional instruction metric may not order architectures in the same way, the reason being that the functional instruction ratio does not consider all of the addressing concepts that may be involved in reducing the space to represent a program or the number of bits moved in execution. It is basically a count of operation code bits.

## 5. ARCHITECTURAL COMPARISONS

Some work has been done, and intense arguments occur, on the issue of the relative efficiency of various register models, addressing formats, and structures. At issue is the efficiency of stack organizations and other forms of register and memory addressing forms.

Using measures of byte space and assumed computer cycle times to perform certain instructions, Cragon has undertaken to compare architectures assuming that the operation code is always 1 byte, the addresses will be 1, 2, or 3 bytes long, and base register referencing will be held within the operation code byte. Several basic architectural models are defined:

1.  *Two-Address Memory-to-Memory.* Number of bytes required per instruction is 1+2B. (B is number of bytes required for address.)
2.  *Three-Address Memory-to-Memory.* Bytes required is 1+3B.
3.  *Accumulator.* Requiring the loading of one operand into an arithmetic register with a nonfunctional instruction and the form LOAD, OPERATION, STORE. The number of bytes required for the entire sequence is 1+B (for operation) + K(1+B). K is the average number of additional instructions. Since not all results will be directly stored and some operands will already be in registers, the LOAD and STORE will not be required every time.
4.  *Register File.* The multiple register machine that performs its operations register-to-register and commonly uses the form LOAD R1; LOAD R2; OP R, R; STORE R1. Bytes required is 2+K(1+B).
5.  *Stack.* Where two stack pushes may be required and the operation requires no address. Bytes required, 1+K(1+B).

Clearly the values of K, the average number of auxiliary or overhead instructions, will have a major impact on the relative efficiency of these architectures. Using studies of code developed from several sources, Cragon establishes some values of K that lead to equations for calculating the byte space requirements for those architectures that depend on values of K, namely, the accumulator, register file, and stack architectures. These equations are as follows:

$$\text{BYTES/ACCUMULATOR:} \quad BA = 1.6(1+B)$$
$$\text{BYTES/REGISTER FILE:} \quad BR = 3.2+1.2B$$
$$\text{BYTES/STACK:} \quad BS = 2.2+1.2B$$

We must remember that these equations are based on studies of coding patterns for various machines. While the data is rather extensive, the measures derived are statistical and we are necessarily dealing with approximations.

By reducing K to a known average, the bytes required for instructions in each architecture become a function of whether 1-, 2-, or 3-byte addressing is to be used.

Making certain assumptions about the number of bytes required by operands; the number of bytes that can be transferred in a single interaction between processor and memory; the times required to perform the functions of an operation; time to apply base displacement addressing values; and so on, a total time for execution of an operation in each architecture is computed.

Cragon concludes that the stack architecture is generally superior; the register file architecture is next. Perhaps the conclusion is less interesting than the assumptions and the approach, which tries to include some design assumptions about periods of time required to perform operations and address formation with the measure of space required as a figure of merit for an architecture. In effect, Cragon is saying that if we hold the position that the design properties of the machines are identical we can measure the architectures in terms of those design assumptions. We will understand the difficulties of this approach when we have finished with Part Two.

The Cragon study shows us that different approaches to architectural merit can be taken and that there are different notions of how pure, how separate from design issues, architectural studies may be. It is interesting that by Cragon's measures, which were very influenced by design assumptions, the pure stack, register file, and accumulator architectures are superior to the two- and three-address storage architecture. It is not only the time measure in which the stack architecture is superior. Using his derived values of K, Cragon indicates that, despite auxiliary instructions, for 2- and 3-byte addresses, the stack and register file architectures will require fewer bytes per operation than the two- or three-address instructions that would minimize necessary LOADs, STOREs, PUSHes, and POPs. Thus Cragon provides an answer, perhaps not *the* answer, to the question of how many nonfunctional instructions one may expect in register-based architectures.

A conclusion that can be drawn is that a pure architecture, or any of the above types, cannot be an optimum architecture. An optimum architecture might provide combinations of various types of register population and addressing model, a mixture of stack, register file, and memory-to-memory operations.


## 6.  IDEAL ARCHITECTURE

The notion of architectural combination and flexibility does indeed occur in one architect's concept of an ideal architecture. This section discusses one idea of an ideal architecture that addresses various problems of work load characteristics by including architectural forms that permit maximum efficiency across various work loads. The discussion is taken from the Flynn articles referenced in the bibliography at the end of this chapter. Flynn provides some rules for an ideal architecture that concern the nature of the instruction set and the nature of addressing and memory referencing. They serve not only to indicate what the desirable properties of an architecture are but to introduce us to some notions of the relation of an architecture to a high-level programming language.

One of the tasks that confront an architect is the determination of exactly what instructions should form the functional instruction set. An answer to this is to provide an operation code for each functional operation in a high-level programming language. Thus operation codes will exist for +, −, *, /, SINE, and so on. Special provision is made for operations on arrays, and all operations that can be applied to data types in a high-level language should be directly applied in the architecture. The instruction set contains only instructions that are used in the higher-level language, there are no nonfunctional instructions, and the notion of a nonfunctional instruction is an instruction that is not expressed as a verb or operator in a higher-level language. Thus LOADS, STORES, and so on are not statements made in high-level languages and therefore represent only architectural overhead.

An instruction word consists of a single operation code and an operand address for each high-level language variable involved in the operation. Operation codes are symmetric in that they are applicable to any type of addressing and any data type. Thus the number of operation codes looks small because there is no designation of operation codes that are dependent on addressing type and data representation.

## 6.1. Addressing and Memory Referencing

We mentioned in Chapter 10 that some architects believe in constrained addressing space to encourage program modularity and structure. In addition, address space requirements should make maximum use of the ability to designate an address by using the structured context in which a program is running to the fullest extent possible.

The size of the address space implies the number of bits that should be used to represent an operand reference in an ideal architecture. It is $\log_2$ of the number of variables that will occur within a subroutine. Thus in an ideal architecture the number of bits required for addressing the workspace of a program is only the number necessary to address within the operand and segment context of a particular module.

As regards the number of required addresses in an instruction, the architecture should provide only as many addresses as are required to describe a specific function.

## 6.2. Characteristics of Ideal Architecture

The above discussion suggests some general criteria for an ideal architecture:

1. Only one instruction should be executed for a high-level language operator.
2. There should be only one memory reference for each referenced operand.
3. There should be explicit addressing only for operands whose location cannot be inferred by recent processing activity, and addresses should be short.

Flynn represents a rather neat structure for the operation codes of an ideal architecture stated in terms of the number of operands required and the relationship between the operands. The structure is oriented toward efficient coding of an instruction set that will support programs written in a higher-level language with minimum compilation.

The instruction set contains 16 fundamental formats of 3, 2, 1, or 0 operand address requirements. There is a stack and a set of stack operations. In general, the forms provide for:

1. An operation on A and B to produce C (three operands).
2. An operation on A and B to produce a new A or B (two operands).
3. An operation on two instances of A to produce a B (two operands).
4. An operation on A and B to produce an intermediate result (put a value on an operand stack).
5. An operation on A and the stack to produce a B (two operands).
6. An operation on A and the stack to produce an A (one operand).
7. An operation on two stack elements (zero operands).

Thus instances of A=A+B, A=B+C, and A=B+C*D are provided for and stack operations are eliminated (the equivalent of eliminating LOADs by eliminating functionally equivalent PUSHes) except where the stack is useful to hold an intermediate result.

The ideal architecture is applied to example FORTRAN statements as follows:

I = X*Y+Z*Z
B = A(I)*B
IF (B) 10,20,30

| | |
|---|---|
| XY* | Multiply X and Y, place result on stack (A OP B to STACK). |
| Z* | Multiply Z by Z (one operand) result to stack (A OP A to STACK). |
| I+ | Add top stack elements result to I (STACK, STACK+1 to A). |
| A(I) | Place A(I) on stack (A OP B to STACK). |
| B* | Multiply top stack by B result to B (A OP STACK to A). |
| IF L | If top of stack < 0, go to 10 (L): if = 0, go to L+1; if > 0, go to L + 2. |

The above instructions achieve the rule of one instruction per high-level language function (if we interpret the formation of the address of A(I) as a high-level language function), minimize the number of addresses required, and may reduce the number of bits required to a minimum depending on the details of operation code and address coding.

## 7.  ARCHITECTURE, COMPILABILITY, AND DESIGN COMPLEXITY

A serious problem that presents itself in measuring architectural efficiency is that most of the programs in the world are generated by compilers. The effective architecture is the set of instructions and programming patterns that are used by the compiler. The instructions generated by the compiler may not be an excellent representation of the potential power of the architecture. Even compilers that have excellent optimizing features and that rearrange code, suppress common subexpressions, optimize register references, move instructions out of DO-loops, and so on, may not discover the best coding patterns for a particular program.

Compiling good programs may be difficult because the architecture is not symmetrical and different coding patterns must be used for different representations of data and different data lengths. Compiling may also be difficult because there are so many options, each of which is best for a special case of flow and data, that a compiler can just not reasonably be expected to analyze all special cases.

Let us look at an example of increasing architectural richness by extending the operation code and the possible impact on a compiler. Postulate that, in a mature architecture, it is noticed that in the call of a subroutine a large number of arithmetic registers must be stored and reloaded on a return from a subroutine. The addition of a STORE MULTIPLE instruction and a LOAD MULTIPLE instruction looks like an excellent architectural extension. In any architecture that is well studied, many opportunities for recognizing various code sequences and introducing more powerful single instructions to perform the same functions may present themselves. It was this possibility that made extendability interesting to the Army-Navy study.

The extension of the instruction, of course, does nothing to improve the behavior of older programs unless they are recompiled or recoded by hand to use the new instructions. The presence of the new instructions, however, requires that the compiler be revised to use them and that programmers still working in assembly language be educated to know when it is proper to use the new instructions.

It would seem at first that the new instructions would be used at all times. But it is the observation of some architects that there are instances where an improved instruction may actually reduce the performance of a machine based on the architecture. For example, if there is only a subset of registers to be stored out of a large population of registers and the STORE MULTIPLE instruction cannot define register subsets, then its execution may move more bits than the sequential execution of a small number of STORE instructions. We have defined a special situation for the compiler and programmer where it is necessary to determine when to use the MULTIPLE instructions and when to use sequences of simple LOAD, STOREs.

To achieve true efficiency a compiler must therefore be extended not only to generate the new instruction, but also to determine when the new instruction should be used. This determination may be model dependent in a way inconvenient to compiler writing.

As an instruction set becomes very rich, the percentage of instructions actually generated by compilers may become rather modest. Various studies indicate that very small subsets of instructions account for a very large percentage of all instructions generated by a compiler. Also addressing forms used by a compiler tend to be standardized and may not represent all of the power of an architecture with a very rich set of addressing form possibilities. Statistics quoted in the literature suggest that as few as one-sixth of the instructions of an instruction set may represent 99% of the instructions actually generated by a compiler for a particular language.

The inference to be drawn from the above is that a very clean and symmetrical instruction set conceived at the beginning of an architectural life is superior to the constant enrichment of an architecture with special-purpose instructions aimed at minimizing instruction count, bit movement counts, or space. An architecture with a minimum number of instructions and address forms is superior to an architecture with a very rich and complex set of instructions and addressing forms.

There are counterarguments, as you may imagine. One is that a general-purpose architecture will support compilers in several languages and that compilers for FORTRAN and COBOL will not use the same instruction set for either the compiler or the compiled programs. Another argument is made that the complexity of an architecture is not well measured by a count of instructions. Proper measures of an architecture have to do with the level of the hardware/software interface, the symmetry of the instruction set in terms of its ability to apply identical functions to all data types and addressing modes, and so on.

Another issue in architectural richness involves organization and implementation. We mentioned above that the idea of extending an architecture was attractive to people compiling the Army–Navy study but that some architects would find this feature undesirable because it would lead to increased architectural complexity, which would in turn lead directly to increased design complexity.

More instruction codes and addressing forms may require more specialized circuits, more space to represent the architecture in physical packaging, and so on, reducing the price/performance of particular models. This is particularly ineffective if the large population of instructions are not going to contribute to efficiency because they are not going to be adequately used by compilers.

There are some very interesting architecture and design trade-offs. Some implementations involve the use of a control memory holding microcode instructions to implement the architecture of an apparent target machine (discussed in Chapter 21). Each instruction on the apparent architecture is performed by the execution of a set of lower-level instructions that effectively simulate wiring for a particular target machine instruction. There is great similarity here with the concept of macroinstructions or pseudoinstructions presented to an assembler that generates real machine instructions to perform the macro.

The definition of a very rich architecture with a large population of operation codes and addressing forms may lead to minimization of the space required to represent a program (assuming a splendid compiler). However, this happens only at the cost of having more microde to implement the rich architecture and the consequent increase in control memory space for those machines using microcode to implement the instruction set. Sometimes the control memory that holds the microcode uses the same or very similar technology as the general-purpose memory does, so that all that changes is the apparent size of a program and the total amount of memory required to execute a program is such that no increase in systems efficiency is obtained.

The question of whether it is then better to have the richer instruction set depends on some notions that seem to constantly contend with each other. A designer might take the position that it is desirable to reduce the instruction set because implementing it will increase the cost of the machine and present some price/performance problems. A pure hearted architect might argue that it is necessary to implement a rich architecture because that will establish the architecture and eliminate the need for later modifications. Also architectures are longer-lived than designs and technology, and some future implementation of the architecture may be undertaken without the disruption implied by late architectural extension.

## 8.  THE GREAT DEBATE

Perhaps a correct impression is that measurements of architectural power are closer in nature to the social sciences than to the physical sciences. Variations in work load assumptions, the subtle interposition of design ideas into architectural speculation, and the changing of implementation techniques make measures of architectural power a very difficult area in which to tread.

This is nowhere made more obvious than in the great debate that has raged in the pages of *Computer Architecture News* from the mid-1970s until the time of this writing. A continuing set of articles on such issues as usefulness of stacks and need for registers has appeared as architects of considerable reputation disagree with each other's data and assumptions. The reader interested in further exploration of the area of architectural comparison and power can get an idea of the state of the art by reading the articles listed under the subheading "The Great Debate" in the bibliography at the end of the chapter.

## 9.  LEVEL OF AN ARCHITECTURE

A currently heated argument concerns the *level* of an architecture. Level conveys the degree to which:

1.  The hardware/software interface should be designed to incorporate into hardware various structural concepts and functions associated with operating systems.

2. The extent to which concepts associated with high-level programming languages should be reflected in the instruction set, control mechanisms, and addressing design of the hardware of the system.

Advocates of extending the hardware architecture to include functions and structures associated with operating systems and to move the hardware architecture closer to the structure of programs written in high-level languages see the following benefits in extension:

1. A reduction of the effort required to debug programs by providing at running time the capability of discovering semantic errors that a compiler cannot discover. For example, the passing of parameters of a mode not expected by the subroutine, the exceeding of array bounds, and references to objects that no longer exist.

2. The improvement of the space required to represent a program by providing very powerful instructions that represent the verbs and operators of a high-level language directly.

3. The improvement of the efficiency of running programs by eliminating instructions that are merely nonfunctional architectural overhead.

4. The reduction of the difficulty of writing compilers by presenting a target architecture that is closer to the structural concepts of the high-level language.

5. The provision of additional security and integrity mechanisms by making hardware responsible for detecting illegal references.

6. Increased efficiency of programming by providing an architecture that encourages modularity.

7. A reduction of the extent to which the system must depend on systems software and therefore increased reliability and performance of a machine.

8. A reduction of the total systems cost by replacing expensive software by inexpensive hardware.

In effect, the advocates of higher-level architectures feel that it is better to rely less on compiler development and operating system development to provide a machine with certain functions.

A counterposition claims that an increase in computer architectural and design efficiency may be achieved by actually reducing the level of the architecture, eliminating a number of instructions and increasing dependence on sophisticated compilers.

Clearly these two schools disagree on a number of points:

1. The actual cost of hardware and whether it is truly cost-effective to represent complex algorithms in any hardware form.

2. The state of the art in compiler design and the extent to which truly efficient compilers can be created at reasonable cost.

3. The extent to which code compaction will actually increase the effective performance of a machine.

One may expect that there will in fact be a movement toward placing more function in hardware and that instruction sets and addressing conventions will in fact move in the direction of functional and addressing concepts found in high-level languages. Whether these trends will result in the development of truly high-level language machines, in the sense that advocates of these machines use the phrase, remains to be seen.

### 9.1. Support of an Operating System

Over the years a set of basic structures and functions associated with operating systems has been discovered. The need for representing the status of a program by process control blocks, for protecting areas of memory, and for formal mechanisms for resource acquisition and release are generally recognized as basic needs of an operating system. Recognition of the needs of the operating system has already resulted in enhanced hardware to support paging and segmentation, in the provision of specialized registers for operating systems use, in the provision of hardware memory protection mechanisms, and in the definition of various control states that provide for differing resource manipulation and access rights.

In addition, certain instructions have been developed that make task switch more efficient, make switching between control states more efficient, and so on. Some contemporary architectures, among them the VAX-11, represent the concept of the process in a defined hardware structure and provide instructions for very efficient saving and restoring, not only of registers but of other information needed to control process execution.

There is current interest in extending hardware to provide even more support for operating systems' function and structure. Instructions to efficiently manipulate lists and queues, fundamental objects manipulated by an operating system, are appearing in architectural papers. Beyond this, many of the operations of an operating system are being provided in specialized instructions rather than in traditional software. Frequently, a set of operating systems functions like interrupt handling, task switching, I/O services, and memory management are provided in microcode. Future technology may provide these functions in very high-density electronic circuitry.

One of the issues in extending hardware to offer more support for operating systems is the determination of what the basic functions and mechanisms of an operating system truly are. For example, a distinction is made in the INTEL 432 and in other places between *mechanism* and *policy*. A mechanism is a basic activity that may be used to implement various algorithms that represent a policy. For example, the basic function of task switch, to save the state space of a program in memory and bring the state space of another program to the processor, is a mechanism. The criteria used to determine the priority of contending programs, to order dispatching lists, are examples of policy.

Another example of policy vs. mechanism lies in the area of security and protection. Protection is a mechanism in the sense that an architecture must have some basic features of memory protection and control state to enforce any statement of who can access what resources of the system. Security is a statement of the policy that determines who can access resources and what rights to read, write, destroy, and so on, a resource any particular user or program may have.

In implementing operating systems in some form of hardware it is necessary to distinguish policy from mechanism. Since policies may be installation and user specific and since there is a great deal of instability in the resource management policies over the life of an operating system, it is better not to bring policies into hardware but to implement only those mechanisms that are judged to be basic and universal for all users of a system.

## 9.2.  Extension Toward Higher-Level Languages

The fundamental argument about the level of an architecture is addressed primarily to this area. Most architects agree that more efficient interfaces between hardware and programs in the operating system area are desirable. Fewer agree on the extent to which the architecture should be extended to make a machine look more like a programming language.

Many argue that the basic instruction sets of contemporary commercial processors and the associated addressing concepts are too far from the notions of program and data structure associated with modern programming languages and their compilers. There is a *semantic gap* that places too much burden on a compiler, makes debugging difficult, makes some errors essentially undetectable, and reduces performance because the compiler must produce essentially inefficient code. This is particularly true of compilers that are directed to produce code that will check for certain common programming errors such as array range exceeded or refer to a decimal value with instructions intended for binary values.

Advocates of higher-level architectures take the position that it is worthwhile to invest in extra circuitry to support a higher-level machine architecture that might even lead to increased performance because of increased compiler efficiency. It would certainly lead to increased program reliability and programmer productivity.

An important contributor to architectural research is Dr. Glen Myers. Myers has attempted to define a machine that would reduce the number of undetected semantic errors, increase security, encourage modularity, improve debugging, and so on. Many features of this architecture are typical of higher-level machine approaches:

1. *The Use of Self-Describing Data.* Each data element is described in a *tag* associated with the element. The tag provides a statement of the data type, the intended use of the element, and so on. There are formal descriptor words for nonscalar elements that provide for addressing elements of an array, for example, through the descriptor words. Among the commercially available machines, many Burroughs architectures use such techniques. The purpose is to allow hardware to recognize when a program is trying to execute a data word or to modify an instruction. The machine itself will raise an error signal when such events occur. In addition, Myers proposes that a called subroutine be provided with an instruction, ACTIVATE, that checks to see that the tags of parameters being passed are in accordance with the tags that the subroutine expects.

2. *Object-Oriented Addressing and Resource Management.* The architecture views a program as a formal structure containing an identifier, a coding portion, a private address space, and a set of capabilities. A capability, referred to in Chapter 10, is a pointer to an object and statement of rights. An address is considered to have two dimensions, the name of an object and an index into the object. Everything in the system is an object of some type. To address outside of the private workspace, a program must receive or request a capability to an object to which it wishes reference. The rights statement in the capability determines what a program can do to a particular object. The architecture contains a set of instructions for the creation, deletion, and control of objects and capabilities. The private address space is constrained to a size that will force modularity for large program development. Addressing beyond the private workspace is entirely through capabilities to external objects.

3. *Very High-Level Instructions.* Instructions are provided for the management of objects. The instruction set is symmetrical with a fully defined set of addressing and mode type faults when operations that are illegal in a high-level language are attempted or when reference to an improper location is made. A set of formal error descriptors is defined for quick analysis of error conditions.

This architecture is not intended to provide a directly executable interface to a high-level language but to provide compilers with an interface that will trivialize the compilation process, remove the requirement for generating instructions to recognize certain kinds of errors from the compilation process, and reduce or eliminate the need for complex compiler optimization. We have alluded to it here as one direction in which those who believe in closing the semantic gap are moving. Some of its features have been seen in part in earlier architectures, some of its features are unique.

One of the problems of extending an architecture in the direction of higher-level languages is the fact that there is a rather large set of higher-level languages, with some obvious and some subtle differences, in use. FORTRAN, COBOL,

PL/1, ALGOL, BASIC, PASCAL, ADA, and APL are all possible candidates for more direct machine support. There are so-called APL, PASCAL, and ADA machines available. The Burroughs 1700 and 1800 have attempted to provide enhanced interfaces for both ALGOL and COBOL by the use of microcode that modifies the architecture depending on which compiler is running or which compiler produced the code being run.

There are architects who believe that a higher-level language architecture does not aim at reducing compilation but at eliminating it so that programs written in higher-level languages are directly executable. Many design notions intrude themselves in descriptions of such approaches because authors are concerned with showing that such an approach can perform well. Thus one concept of a high-level machine includes a lexical processor that runs autonomously to translate strings of higher-level language into usable control and address tokens that are passed onto execution processors that execute in parallel with the interpretation and translation of successor high-level language statements.

## QUESTIONS

1. Explain the notion of the power of an architecture as distinct from the notion of the power of a machine model.

2. What is a basic metric for architectural power?

3. What is a functional instruction?

4. What are three characteristics of an "ideal instruction set"?

5. What benefits are claimed by proponents of high-level architectures?

6. Discuss the distinction between mechanism and policy and how it affects the machine/operating system interface.

7. What is self-describing data?

8. What are some characteristics that are appropriate for higher-level language machines?

## BIBLIOGRAPHY

This bibliography provides source material for Chapter 12.

*Computer Architecture News* is represented as *CAN*; *Communications of the Association for Computing Machinery* is *CACM*, *Proceedings of ACM Symposia on Computer Architecture* are represented as *ACM-SIGARCH*.

Burr, W. E., and W. R. Smith. Comparing Computer Architectures, *DATAMATION*, Vol. 23, No. 2, February 1977, p. 48.

Chu, Y. Direct-Execution Computer Architecture, *CAN*, Vol. 6, No. 5, December 1977, p. 18.

Clark, D. W., and W. D. Strecker. Comments on 'The Case for the Reduced Instruction Set Computer,' *CAN*, Vol. 8, No. 6, 15 October 1980, p. 34.

Cragon, H. G. An Evaluation of Code Space Requirements and Performance of Various Architectures, *CAN*, Vol. 7, No. 5, 15 February 1979, p. 5.

Dennis, J. B. Computer Architecture and the Cost of Software, *CAN*, Vol. 5, No. 1, April 1976, p. 17.

Dennis, J. B. Research Directions in Computer Architecture, National Technical Information Service, AD/A-061 222, September 1978.

Ditzel, D. R., and D. A. Patterson. Retrospective on High-Level Language Computer Architecture, ACM-SIGARCH 7th Annual Symposium on Computer Architecture, May 1980, *SIGARCH Newsletter*, Vol. 8, No. 3, p. 97.

Flynn, M. J. A Canonic Interpretive Program Form for Measuring 'Ideal' HLL Architecture, *CAN*, Vol. 6, No. 8, April 1978, p. 6.

Flynn, M. J. Directions and Issues in Architecture and Language, *IEEE Computer*, October 1980, p. 5.

Fuller, S. H., H. Stone, and W. E. Burr. Initial Selection and Screening of the CFA Candidate Computer Architectures, *AFIPS Conference Proceedings*, National Computer Conference 1977, Vol. 46, p. 139.

Myers, G. J., Storage Concepts in a Software-Reliability-Directed Computer Architecture, *ACM-SIGARCH, 5th Annual Symposium on Computer Architecture*, p. 107.

Myers, G. J., and B. R. S. Buckingham. A Hardware Implementation of Capability-Based Addressing, *CAN*, Vol. 8, No. 6, 15 October 1980, p. 12.

Mountain, J. B., and P. H. Enslow. Application of the Military Computer Family Architecture Selection Criteria to the Prime 400, *CAN*, Vol. 6, No. 6, February 1978, p. 3.

Patterson, D. A., and D. R. Ditzel. The Case for the Reduced Instruction Set Computer, *CAN*, Vol. 8, No. 6, 15 October 1980, p. 25. Peuto, B. L., and L. J. Shustek. An Instruction Timing Model of CPU Performance, Proceedings 4th Annual Symposium on Computer Architecture, *CAN*, Vol. 5, No. 7, March 1977, p. 165.

Rattner, J., and G. Cox, Object-Based Computer Architecture, *CAN*, Vol. 8, No. 6, 15 October 1980, p. 4.

Stevenson, J. W., and A. S. Tannenbaum. Efficient Encoding of Machine Instructions, *Can*, Vol. 7, No. 8, 15 June 1979, p. 10.

Tannenbaum, A. S. Implications of Structural Programming for Machine Architecture, *CACM*, Vol. 21, No. 3, March 1978, p. 237.

## The Great Debate

Keedy, J. L. On the Use of Stacks in the Evaluation of Expressions, *CAN*, Vol. 6, No. 6, 1978, p. 22.

Keedy, J. L. On the Evaluation of Expressions Using Accumulator Stacks and Store-to-Store Instructions, *CAN*, Vol. 7, No. 4, 15 December 1978, p. 24.

Keedy, J. L. More on the Use of Stacks in the Evaluation of Expressions, *CAN*, Vol. 7, No. 8, 15 June 1979, p. 18.

Myers, G. J. The Case Against Stack-Oriented Instruction Sets, *CAN*, Vol. 6, No. 3, August 1977, p. 7.

Myers, G. J. The Evaluation of Expressions in a Storage-to-Storage Architecture, *CAN*, Vol. 6, No. 9, 1978, p. 20.

Schulthess, P. T., and E. P. Mumprecht. Reply to the Case Against Stack-Oriented Instruction Sets, *CAN*, Vol. 6, No. 5, December 1977, p. 24.

# PART TWO

# ORGANIZATION AND IMPLEMENTATION

# Concepts in Organization and Implementation

## 1. INTRODUCTION

The organization and implementation of a computer system involves the translation of the architecture into functional units that are ultimately represented in an underlying electronic technology.

The representation of an architecture in hardware is constrained by the cost/performance goals of a system or a family of systems. These goals determine the nature of the basic electronic technology that will be used in the construction of a system. They also influence the number of circuits a system will have and the way that the circuits will be used. Costs are partially a function of the speed of the electronic technology, its power, cooling, packaging, and reliability characteristics, and partially a function of the organization of the machine. By organization we mean the degree of parallel operations, the extent of overlapped operation, the extent of the use of uniquely designed components, and so on.

There are degrees of freedom in the mapping of architecture into organization and organization into physical units. Thus a designer can look for very inexpensive representations of the architecture that may not be fast performers. He can look for very fast implementations of the architecture that may be expensive because a great deal of additional organization is necessary to achieve a certain speed from a given architecture. He can look for optimum price/performance representations of an architecture that exist at some knee determined by available technology.

The concept of the knee suggests that, at any particular technology time period, there are organizations that can be made faster for relatively small increments in organization and component expenses. There is an optimum mapping of an architecture, and there is a set of organizations and implementations that increases the speed of the architecture, but at a point of diminishing return where disproportionate expense is experienced for incremental improvement in system performance.

The degrees of freedom permitted in the organization and construction of an architecture permit families of machines to be built that have the same architecture but offer different price/performance ratios. A "family" means starting with a very cheap but somewhat price inefficient model and progressing through to very fast and expensive price inefficient models by way of good price/performance systems that are neither very small nor supercomputers.

## 2.  ORGANIZATION AND IMPLEMENTATION DECISIONS

The organization and implementation involve a large number of interrelated decisions about how to design the electronic components that will perform in accordance with the architecture. Decisions about the nature and population of electronic components must be made within a framework of architectural requirement, available technology, and price/performance goals.

### 2.1.  Width of Data Paths

This is the number of bits that will flow from place to place in the processor in parallel. For example, the transmission of a 16-bit word from one register to another may be done serially 1 bit at a time, in perfect parallel 16 bits at a time, or in various groupings of 4 or 8. Given a certain physical distance from one register to another, and a particular technology, the amount of time required to move 16 bits between them is a function of how many bits are moved in parallel. Moving a 16-bit word serially 1 bit at a time will take considerably more time than moving 16 bits at the same time in parallel. Parallel movement is faster but costs more because there must be a unique path for each bit moved in parallel and some circuitry to control skew.

### 2.2.  Degree of Circuit Sharing

This is the degree to which circuits will be shared across different functions, requiring that these functions be performed sequentially even though they are not logically related. There are various functions in the execution of an instruction or in the execution of series of instructions that may be performed in parallel. For example, the actual addition for an ADD instruction can be performed while the final address for a successor STORE instruction is being computed. This logical independence can be supported by providing unique hardware that can operate in parallel for all of the logically nonsequential operations. However, if circuits are shared between functions they become dependent on access to the physical circuitry and must operate sequentially even if they aren't logically related. The duplication of circuits to permit this overlap of function will make a machine more expensive but will increase its speed.

## 2.3.  Definition of Specialized Units

This is the degree to which functionally specialized units that allow a particular function to be performed with disproportionate speed will be used. Some instructions require considerably more time to execute than others. For example, a DIVIDE FLOATING POINT takes more time than a LOAD. It is possible to invest in specialized floating point divide execution units in a system so that this operation is performed disproportionately quickly. The success of this investment is determined by the relative frequency of use of the floating point operation and the relative cost of the special circuitry.

## 2.4.  Parallelism of Functional Units

This refers to the degree to which specialized instruction units will be used to allow instructions to be overlapped with each other, gaining speed by that overlap rather than by special functional speed. A characteristic of highly parallel machines is to undertake to execute unrelated instructions at the same time. Certain instruction sequences permit some instructions to actually be executed at the same time so that the elapsed time to perform a set of instructions may be reduced to the elapsed time required to perform one of them. Speed is achieved by this overlap rather than by forcing the organization and technology of one unit to be particularly fast.

## 2.5.  Buffering and Queuing

This refers to the use of specialized buffering and/or enqueuing areas to provide more efficiently accessible local storage to functional units. We shall see that, for faster processors, the speed of logic and the ability to move bits from one place to another in a processor or to modify bit streams by logical and arithmetic operations is higher than the speed with which large memories can deliver data to a processor or accept data from a processor. This speed imbalance is addressed in some organizations by the use of relatively small and fast memories in the processor to hold instructions that the processor will soon execute or to hold data that the execution elements require.

## 2.6.  Prediction

Prediction is the degree to which special attention is paid to predicting the functions that will be performed in the near future by projecting the behavior of the system in the recent past. This involves the collection, storage, and manipulation of data in special tables in hardware. The intent is to try to guess what the processor will be doing in the near future and optimize data movement based on those guesses. An example is a machine that remembers whether or not a particular branch instruction has caused transfer at its last execution and to stage the fallthrough or transfer instructions forward to the processor when the branch is encountered again.

Predictive techniques are often used for the control of the contents of small buffer memories in order to improve the probability that what is brought to the buffer will actually be used.

## 2.7.  Underlying Technology

This is the speed of the underlying technology that will be used to implement functions. There are various technologies available for the manufacture of processors, memories, and other components of a computing system. Thus a machine of a certain speed can be built with relatively simpler organization if faster technology is used than if slower technology is used.

## 2.8.  Functional Implementation

This refers to the degree to which functions are represented in logic organization and fast circuits, as opposed to other types of representation. An organization may call for the representation of function in a variety of ways. Functions may be hard-wired so that a particular operation is built into the machine. They may be represented by patterns of bits in a special control store, each bit representing some condition of the circuits that should be set to perform a certain function. Functions may also be represented in programs written in the architecture of a smaller computer imbedded in the larger computer. The instructions of the larger computer are executed by subroutines in the smaller computer that effectively simulate the logic of the larger computer.

## 3.  PRICE/PERFORMANCE GOALS

The price/performance of a system is computed by dividing the price of a configuration of a system into some statement of the power of the system. Thus one achieves a price/performance ratio by adding the price of processors, memories, storage devices, terminals, and so on to determine a systems cost and dividing this into some statement of systems power.

The major difficulty in this procedure is developing a statement of systems power that is meaningful and general. We have all observed that some machines do certain things better than others, and we have all observed that different applications have different work load characteristics. It is frequently necessary to constrain the notion of price performance to a particular application so that transactions in that application can be characterized, and we can talk meaningfully of so many transactions per second for a particular configuration.

Computer designers of general-purpose systems face the difficulty of assessing the price/performance of the systems across a number of various configurations and across a very generalized notion of work load. The search for some common denominator of systems performance is very complex.

Since the cost of storage devices, terminals, and other peripheral equipment is to some extent independent of the price of the processor, it is convenient to attempt a figure of merit for the processor alone. Figures of merit for processors are necessarily approximations that take into account the speed of particular instructions adjusted by the probability of use of various instructions in a program.

With a notion that X% of instructions will be types of branching instructions, Y% will be data movement instructions, and so on, one can come up with an average instruction time. This time can be converted into the notion of the number of instructions the processor will execute over a given period of time. A common representation of this notion is the idea of MIPS or KIPS. MIPS represents millions of instructions per second, KIPS represents thousands of instructions per second.

Thus a figure of merit for a machine is the MIPS rate divided into:

1.  Price for the processor; Plus,
2.  price for the amount of memory required to hold enough data and instructions so that the processor can execute instructions at its maximum rate; Plus,
3.  price for the interconnection to enough data devices to allow the flow of data through the machine with sufficient speed so that the memory always has sufficient data and instructions for meaningful processing.

## 4.  ARCHITECTURE, ORGANIZATION, AND PRICE/PERFORMANCE

The organization and implementation of a machine at a certain price/performance level is particularly difficult work for very small and very large versions. The challenge for the very small version is the discovery of the minimum organization and equipment necessary to achieve a given level of performance. The challenge at the top of the line is the discovery of those parts of the organization that will have maximum performance impact assuming various patterns of instruction use.

In different environments, facing different work loads, processors will execute different combinations of instruction sequences. In order to maximize performance at reasonable cost, various assessments must be made of the percentage of time a processor will be executing particular instructions, the probability that sets of instructions can be executed independently of other instructions, and so on. If, for example, the machine is expected to be used primarily for scientific calculations involving floating point operations, then the floating point instructions can be made particularly fast relative to the rest of the other instructions. Or it is possible to devise ways of executing floating point instructions in parallel with other unrelated instructions. If guesses about the instruction sequences are not correct, however, and the processor encounters programs that do not use the expected instruction patterns, the performance of the machine for those programs will be disappointing and the machine will not be considered a general price/performance success.

The architecture of the processor will affect how much organization is necessary for high performance in a number of ways. The architecture will determine the probability that consecutive instructions can be executed in parallel and how much *lookahead* and *staging* must be done to execute an instruction or group of instructions quickly. The mechanisms for achieving lookahead and instruction staging are discussed in detail in Chapters 16 and 17. The fundamental notion is that a processor organization may have the ability to acquire instructions beyond the instruction it is currently executing, begin to prepare these instructions for execution, and, possibly, execute more than one instruction at a time. For example, the instruction sequence

1.   LOAD R1, A
2.   ADD R1, B
3.   STORE R1, C
4.   LOAD R2, D
5.   MPY R2, E
6.   STORE R2, F

could execute the addition and the multiply sequences in parallel. This is due to the fact that there is no logical ordering relationship between them and, consequently, the R1-oriented and R2-oriented instructions are commutative. If lists of instructions are commutative they may be executed in parallel in an organization that provides enough circuitry. The architectural potential for parallel execution comes from the ability to isolate the arithmetic in separate registers. The organizational potential comes from the provision of enough circuits to execute multiple LOAD instructions in parallel and by the provision of separate ADD and MULTIPLY units.

Notice the following sequence in a different architecture:

1.   ADD A, B, C
2.   MPY D, E, F

For the first machine to find instructions to execute in parallel it would have to look past the STORE R1 for sequences of logically unrelated instructions, while the second machine, because of denser packing of each instruction, might characteristically find consecutive instructions executable in parallel. Less circuitry and intermediate instruction storage space would be required in the three address architectures to accomplish lookahead resulting in parallel execution. A fast version of the second architecture, however, would need an ability to form addresses in parallel very quickly since the speed of the machine is so relatively dependent on its ability to develop three addresses for each instruction in minimum time.

When we talked about the architectural efficiency of these instruction forms we were concerned about program compaction and bit flow. The designer is concerned with parallel execution capability and the cost in complexity of design and circuitry to achieve various levels of execution speed in various models of the architecture.

The designer looking for price/performance must evaluate how best to use circuits. Is it better, in a technology period, to use the circuits to support more sophisticated architectures, to support mechanisms for more speed for less sophisticated architectures, or to risk a more aggressive technology so that more speed can be achieved with simpler organizations?

Within a broad range of performance goals, different architectures may be organized to achieve roughly similar price/performance goals. However, in certain special-purpose environments, among them very high-speed scientific computing, there are points of diminishing returns. That is, the organizational complexity necessary to achieve certain performance from the architecture is so great that price/performance cannot be achieved relative to the price/performance of a machine with an architecture more suited to high-speed computation.

Architecture affects organization, then, in a very basic way. It determines how complex the organization must be to achieve speed with the instruction set, register population, and addressing forms, and with the coding patterns that emerge as a result of the architecture.


### QUESTIONS

1.  What is the notion of a price/performance knee?

2.  What impact will moving more bits at the same time have on performance?

3.  What is the role of a functionally specialized unit in making a processor faster?

4.  What is a rough way of calculating price/performance?

5.  What impact does architecture have on organization?

# Chapter 14

# Basic Concepts of Instruction Execution

## 1. BASIC CYCLES

The fundamental speed of a machine is determined by the speed at which its technology operates. This fundamental rate is determined for a technology by its *basic switching speed* and its *circuit density*. Basic switching speed is the time it takes to set a basic component in the technology from on to off or off to on. Circuit density determines the distance basic components are from each other and, consequently, the amount of time it takes to send a signal from one component to another. Since electronic signals travel at a fixed rate, the closer basic components are the faster they can exchange electronic signals.

The density of a technology determines how many basic components may exist within a space and, consequently, the probable distances that must be traversed between different electronic components. The density determines how many components may be placed on a single *chip*. Electronic components that are on the same chip communicate with each other in minimum time. Not all of the electronics of a computer may be fit onto the same chip, however, so that some signals must move between chips on the same *card*, a packaging vehicle for chips, and some signals must go across cards, or even across *boards*, a higher-level packaging for cards. As signals must cross chip, card, and board boundaries, communication slows due to the distance that a signal must travel. This slowing is due to *propagation delay time*. This is a statement of how long it takes a signal to move from one component to another as a result of the physical distance between them. When a computer is designed, packaging decisions must be made that will group circuits and functions most efficiently in order to minimize propagation delays. The farther apart elements are, the more boundaries must be crossed and the longer a particular function will take.

The underlying technology of a machine determines the duration of an interval called a *basic logic cycle*. A basic logic cycle is the amount of time required to change the state of a logical element of the machine. A basic logic cycle is determined by switching time and propagation delay time.

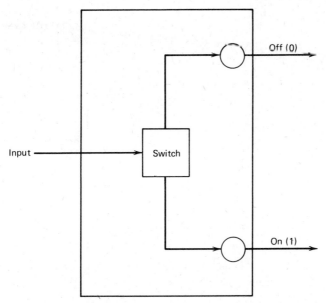

**Figure 5.** Simple flip-flop.

Consider Figure 5. The device pictured is called a *flip flop*, a circuit that changes its state each time an input signal is received. The flip flop is emitting either an on or an off signal as output at any time. When it receives an input signal it changes its condition.

Within the context of various machine functions we may want to set the flip flop to plus or minus (on or off) and use the emitted signal to control the flow of data in the machine. The time required to set or reset the flip flop in the technology of the machine is its basic logic cycle.

Basic logic cycles are organized into a larger cycle called the *basic machine cycle*. A basic machine cycle is an organization concept describing a series of basic logic cycles that are combined together to accomplish a function meaningful in the data flow of a machine.

Consider Figure 6. It contains a register A and a register B, each holding 1 bit. We consider the basic machine cycle to be the interval, from time $t_0$ to $t_3$, that it takes to transfer a bit from register A to register B.

The interval from $t_0$ to $t_3$ consists of three basic logic cycles required to accomplish the transfer of data. On logic cycle 1 a signal is sent to register A causing it to open its output gates. On logic cycle 2 the contents of register A are put on the output line. On logic cycle 3 the contents of register A arrive at register B.

From this point on we discuss the speed of a machine in terms of the times associated with the basic machine cycle. The basic machine cycle is considered the elemental unit of time. We make the simplifying assumption that a basic machine cycle always contains the same number of logic cycles so that it is

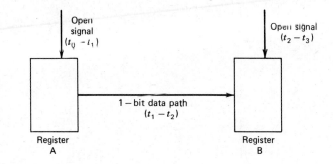

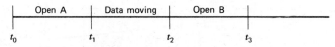

**Figure 6.** Basic machine cycle.

always the same regardless of the specific functions being performed during the cycle. The speed of a machine, therefore, varies in terms of the number of basic machine cycles that are involved in carrying out specific instructions and control operations common to all instructions.

The duration of a basic machine cycle is determined by the duration of a basic logic cycle, the number of logic cycles involved in a machine cycle, and the extent to which basic logic cycles may be performed in parallel during a machine cycle.

The machine cycle may involve parallel operation in two ways. First, more than 1 bit may be moved from A to B at the same time. This is known as increasing the width of the *data path* of the machine. Figure 7 shows an 8-bit wide path from A to B, providing for an 8-bit transfer during a machine cycle interval $T_0$ to $T_1$. (Basic logic cycles are shown below the machine cycle interval.)

Another way of using parallel function in a machine is to increase the number of distinct functions that are going on at the same time. Thus, while 8 bits are moving from A to B, 8 other bits may be moving from D to E, and a single-bit signal may be moving from G to H, and so on.


## 2.   EXECUTION OF AN INSTRUCTION—INSTRUCTION STAGES

A particular instruction on a machine will involve a great deal of data movement in the logic while the instruction is being interpreted and executed. The speed of an instruction is determined by duration of the machine cycle and the number of sequential machine cycles required to execute the instruction.

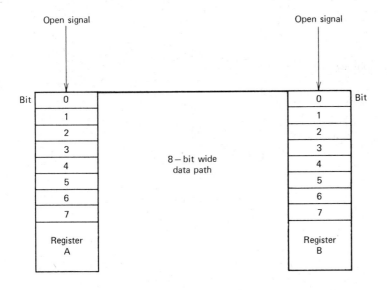

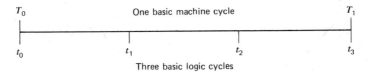

**Figure 7.**  Parallel transfer.

Consider the processor represented in Figure 8. The figure shows a very simple organization supporting a simple architecture for the purpose of this discussion. The boxes in the figure represent machine registers. The numbers in the boxes indicate the width in bits of the register. Only one box, computer control circuits, is not a register. It is a collection of circuitry and additional registers that aren't indicated. The lines in the figure represent paths for data to flow from one register to another. Control circuits and signal paths that may be 1-bit wide are not shown. The arrows on the lines indicate the direction of flow. When a path to a point is shared, indicated by an intersection of lines, data may not move in parallel from the points involved. Thus we cannot move data from the *memory interface register* (*MIR*) to the *base register* (*BR*) and to the *arithmetic register* (*AR*) at the same time. We could, however, move data between the MIR and the BR and between the MIR and the IR at the same time because there are separate data paths between MIR and IR and between MIR and BR. Since MIR holds only 16 bits, this could be done in this figure only if MIR were transmitting the same value to IR and BR in parallel.

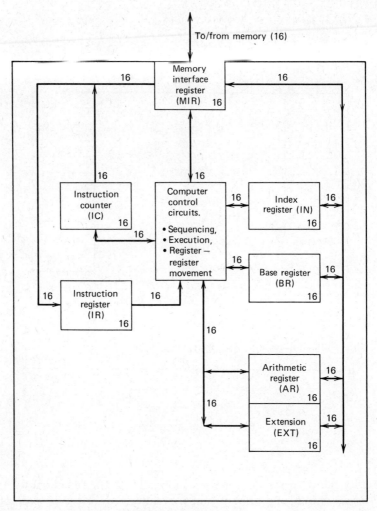

**Figure 8.**   Simple processor.

In Figure 8, as in others of its type that follow, we undertake maximum simplification and informality in representing the lines indicating all possible parallel connections. In this way we hope to simplify drawings. In addition, we only show relevant lines and do not represent all of the connections that might be necessary for a complete organization of the architecture.

The processor shown has only one arithmetic register (AR), one index register (IN), an instruction counter (IC), an instruction register (IR), and a memory interface register (MIR). It also has a base register (BR). The processor architecture defines 16-bit words as the basic unit of manipulation. All instructions are 16 bits and all data words are 16-bits. All registers are 16 bits in length. The breakdown of an instruction is 6 bits of operation code, 1 bit to designate the use of the index register, and 9 bits of address. The base register is always used to form an address.

The registers in the architecture (visible to a programmer or compiler) are the AR, the IN, the IC, and the BR. The BR is referenced only through the use of a LOAD BR instruction that is not protected by control state. The IR and MIR are in the organization but cannot be referenced by program.

The figure also shows a black box representing the control and arithmetic circuits. These circuits contain logic for the manipulation of the registers in the forming of addresses and obtaining and operating on operands. Details are obviously not shown. We ascribe characteristics to this black box as we need them. The basic function of the box is to control the flow of data between registers and to place and take values from registers for operations to be performed on them.

The flow of data in the processor of Figure 8 is summarized in the following table:

Register Flow

| | |
|---|---|
| IC: | To MIR. Request instruction. |
| MIR: | To IR. Provide instruction. |
| | To IN. Place index value. |
| | To BR. Place base value. |
| | To AR. Place operand. |
| | To EXT. Place operand. |
| | From IN. Store index value. |
| | From BR. Store base value. |
| | From AR. Store operand. |
| | From EXT. Store operand. |

Control Flow

| | | |
|---|---|---|
| Control circuits | To IC: | Incremented address of instruction. Branch address. |
| | To IN: | Index modification. |
| | To AR and EXT: | Operands. |
| | To MIR: | Operand address. |
| | From IN: | Index value. |
| | From BR: | Base value. |
| | From IC: | Instruction address. |
| | From IR: | Instruction. |
| | From AR,EXT: | Operands. |
| | From MIR: | Operands from memory. |

Let us consider the execution of an instruction on this machine. The instruction LOAD M, IX (IX indicates that indexing is to be used) places in the AR the contents of the memory location designated by the addition of BR, IN, and M (the 9-bit address in the instruction). The execution of this instruction involves the sequence of machine functions shown below. The parentheses are to be read as the phrase "contents of" so that (IC) is read as "the contents of the instruction counter."

1.  (IC) to MIR              Request LOAD instruction from memory.
2.  (MEMORY) to MIR          Instruction to processor.
3.  (MIR) to IR              Instruction to instruction register.
4.  DECODE                   Interpret OP CODE to determine function.
5.  (BR) to ADDER            (BR) to adder input.
6.  M (From IR) to ADDER     Nine bits of address to adder input.
7.  ADD                      Sum of BR and M to adder output.
8.  (IN) to ADDER            Contents of IN to adder input.
9.  ADD                      Sum of BR, M, and IN to adder output.
10. (ADDOUT) to MIR          Address in adder output to MIR.
11. (MEMORY) to MIR          Operand to processor.
12. (MIR) to AR              Operand to arithmetic register.

The execution of the instruction involves 12 sequential stages. The organization of the machine has defined the instruction execution sequence by defining the data flow from the various registers and the sequence in which that data flow will occur. The first four stages, involved in the acquisition and interpretation of an instruction, are commonly referred to as *I-time stages*. The stages following DECODE, concerned with the execution of a specific instruction, are commonly referred to as *E-time stages*.

## 3.  INSTRUCTION STAGES AND MACHINE CYCLES

We would like to relate the sequence of 12 instruction stages shown above to the concept of basic machine cycles. One way of discussing the speed of a machine is in terms of the number of basic machine cycles an instruction requires.

The simplest relationship between machine cycles and instruction stages would be one to one. This 12-stage instruction would then take 12 basic machine cycles.

Since the organization provides for 16-bit wide data paths, there are no additional machine cycles required to move data in smaller chunks. Therefore there is a fundamental possibility that we can approach the execution of a stage in one cycle.

We can observe initially that all movements of data from register to register would take only 1 machine cycle. Those instruction stages that move data from one register to another will each take 1 machine cycle (consisting of a number of basic logic cycles not specified here).

Two difficulties emerge, however, in attempting to relate machine cycles to logical stages on a one-to-one basis in the example machine. The first of these has to do with the memory references. Memory response times, as we discuss later on, are frequently slower than processor logic times. This is especially true of faster processors. As a result of this a number of basic processor cycles may pass before the memory has placed an instruction or an operand in the memory interface register. While it is not universally true that processors are faster than their memories it is a factor to be considered in determining the speed of a processor/memory complex. Many organization and implementation features of more sophisticated processors attempt to address this speed imbalance.

It is possible to organize so that the processor can do some useful work while waiting for a memory reference to be satisfied. Many processors, however, are idle during this time. This will increase the basic machine cycle count when mapping stages against cycles.

The other difficulty in mapping stages on cycles comes as a result of the formation of an operand address involving the addition of (BR), (IN), and the value M from the instruction. This process is shown as instruction stages 5 to 9 in the above list of stages. Stages 5, 6, and 8, which move data from the BR, IN, and IR to inputs of the adder, are 1 machine cycle stage. However, the multiple ADDs involved in the address formation may involve multiple machine cycles, characteristically a cycle to add and a cycle to handle carries.

The example shows that the logical stages of an instruction execution are not directly translatable to basic machine cycles even when the data paths of the processor are identical to the data structures of the architecture. Some of the logical stages involve more than 1 basic machine cycle.

Consider the impact of an organization with only 8-bit wide data paths for the simple processor we have been discussing. Each 16-bit transfer between registers in the processor data flow would involve two sequential transfers of 8 bits each. Two basic machine cycles would be necessary for each register-to-register movement, and each logical instruction stage would involve 2 machine cycles. In addition, some additional control circuits would be necessary to select 8-bit segments from the registers and to select upper and lower half register positions for the receiving registers.

The motivation for 8-bit organization would lie in the anticipation that the savings in the width of the data paths between registers in the data flow of the processor would be greater than the cost of providing some incremental control for the 8-bit movements. The model of the architecture would be slower but cheaper.

Whether the resulting organization would provide a machine that was an acceptable price/performer would depend on the details of the cost of circuits for data paths at the time of implementation. But the ability to make such organization choices demonstrates the degree of freedom between organization and architecture.

## QUESTIONS

1. What is switching time?

2. What is propagation delay?

3. What is a basic logic cycle?

4. What is a basic machine cycle?

5. What is an instruction stage?

6. How do instruction stages relate to basic machine cycles?

7. Why may instructions require more basic machine cycles than logical stages?

# Chapter 15

# Organization for Increased Performance

## 1. ORGANIZATIONAL APPROACHES TO FASTER MACHINES

The previous chapter concluded with some remarks about how to organize a slower version of the simple processor. This chapter begins a discussion of how to make a faster version. In this chapter we investigate how to make the execution of a single instruction faster. There are a number of possible approaches to doing this:

1. Make the underlying technology faster in order to execute the same instruction stages in the same number of machine cycles but with each machine cycle faster. Rather than elaborate the organization of the machine, leave the organization simple and use faster technology.

2. Redesign the execution of instructions to reduce the number of stages by doing more in parallel within a stage.

3. Redesign certain instruction stages to reduce the number of machine cycles they require.

The reorganization can be done in order to improve the times associated with the acquisition, interpretation, and preparation of every instruction. The sequence of stages that are common to all instructions is called *I-time*. Improvements in this area will make the machine faster for every instruction.

Alternatively, the reorganization can be addressed to particular instructions, most likely those that are slowest or most frequently used, in order to minimize the number of machine cycles they consume during the time they are performing their specific functions. This execution time is called *E-time*.

The simple architecture we have been discussing limits some of the organizational possibilities. However, even with the simple example we can demonstrate some of the organization techniques that may be used to make a faster version of the architecture. Richer architectures might naturally lend themselves to more design possibilities. That is a positive way of saying they might also require more organizational effort.

## 2. INSTRUCTION TIMES

Continuing with the LOAD instruction as our basic example, the specific relationship between the instruction stages and the basic machine cycles of the 16-bit organization discussed above is as follows: (We are going to assume that it takes the memory 5 processor cycles to provide data to the MIR. This ratio will vary from system to system.)

| Stage | | Cycles | |
|-------|---|---|---|
| 1. | (IC) to MIR | 1 | Request instruction. |
| 2. | (MEM) to MIR | 5 | Instruction to processor. |
| 3. | (MIR) to IR | 1 | Instruction to instruction counter. |
| 4. | DECODE | 2 | Interpret instruction. |
| 5. | (BR) to ADDER | 1 | Address element to adder. |
| 6. | (IN) to ADDER | 1 | Address element to adder. |
| 7. | ADD | 2 | Sum address elements. |
| 8. | M (from IR) to ADDER | 1 | Address element to adder. |
| 9. | ADD | 2 | Form final address. |
| 10. | (ADDOUT) to MIR | 1 | Request operand. |
| 11. | (MEM) to MIR | 5 | Operand to processor. |
| 12. | (MIR) to AR | 1 | Operand to arithmetic register. |
| Total cycles | | 23 | |

For a general sizing of this machine, let us assume that the basic machine cycle for the processor is 50 nanoseconds ($10^{-9}$ seconds). The elapsed time for the instruction would be 1150 nanoseconds, or 1.15 microseconds ($10^{-6}$).

If we knew what the cycle counts for other instructions were, and if we knew the distribution of instruction executions, we might with some caution estimate the instruction rate for this architecture using the 50-nanosecond basic machine cycle and the stage organization as presented. The characteristics of the machine that we have so far described imply architecture and organization that would place it in the mid-size "minicomputer" range with an execution rate at the level of 600 KIPS (thousands of instructions per second).

## 3. CYCLES REDUCTION BY STAGE SPEEDUP

A first place to look for speed is in the multicycle stages. These are the stages involving memory references, the *decode stage* and the address formation stages.

Decode is the process by which the bit pattern representing the operation code is used to define a hardware state in which certain signals are generated and data paths are opened or closed to achieve a successor machine state or a group of successor states. The specific bit pattern of the operation code will cause a unique pattern of signals to flow through the logic to develop signals that open and close the data paths of the machine and determine the data flow from register to register.

By an elaboration of the organization and an increase in the number of circuits involved in the decode process it is possible, and common on larger machines, to define a 1-cycle decode stage. Since decode is an event that happens to all instructions, such an extension of decode organization is very likely going to be profitable. We discuss some issues in decode a little further along.

The address formation addition is a multicycle stage that involves a first cycle for bit-by-bit addition and a second cycle for bit position carry adjustment. A more elaborate adder would, with increases in logic and circuit count, be able to conduct the addition in 1 cycle.

In an architecture with short addresses in the instruction (so that addition of the contents of a base register is required to gain full addressability to the memory), it is crucial that this address formation addition be done quickly. In order to accomplish this we provide a specialized fast address adder that can form an address in 1 cycle. This unit is not usable for operand addition because it does not perform all of the signed algebraic addition functions. The restriction of the general arithmetic capability minimizes the number of circuits required for the fast address formation.

The speeding of decode and the address formation stages reduces the number of elapsed machine cycles to 20. The only remaining multicycle stages are the memory references. A significant increase in the speed of an instruction may be obtained if a way is found to reduce the effective number of machine cycles required for the movement of data or instructions between processor and memory. We discuss this issue in later chapters. There are still some methods for increasing speed that relate to processor organization alone.

### 4. STAGE REDEFINITION

An inspection of the list of stages reveals an opportunity for achieving an additional increase in speed beyond what is available by adding circuits to speed address formation.

The stage list makes it seem that stages 1, 2, 3, and 4 are sequential and can be neither combined nor executed in parallel. Notice, however, stages 5, 6, and 8:

Stage 5   (BR) to ADDER

Stage 6   M (from IR) to ADDER

Stage 8   (IN) to ADDER

The transfer of the components of the address to the address adder (the contents of IN, BR, and the address portion of IR) could be accomplished in parallel since they are not dependent on each other, and the values can be sent in any order to the adder input circuits. Since there are unique paths from BR, IN, and IR to the input registers of the address adder, no additional path circuitry is needed.

However, we need to give to the address adder the capability of receiving three rather than two inputs and an elaboration of control so that it may receive the three inputs in parallel. Doing this allows us to reduce the time it takes to deliver address components to the address adder from 3 to 1 sequential machine cycles. If we assume an adder organization that accomplishes an addition in only 1 cycle, then the entire elapsed time for the formation of a full address would be 3 rather than 7 machine cycles and would be accomplished in three rather than five instruction stages. The stage redefinition is as follows:

| | |
|---|---|
| Stages 1–4 as before | Reduced to 8 cycles due to 1-cycle decode. |
| Stage 5 IN, BR, M (from IR) to ADDER | One cycle in parallel. |
| Stage 6 ADD IR, BR | One cycle. |
| Stage 7 ADD IN | One cycle. |
| Stages 8–10 (reduced stages) | As before. |

Notice that there is an alternative organization that would move (IR) and (BR) to the address adder in one stage and then define another stage where (IN) was moving to the address adder while the first addition was going on. The elapsed time in machine cycles would still be 3.

## 5. RESEQUENCING DECODE AND ADDRESS FORMATION

An approach to speeding an instruction that requires address formation involves expanding our notions of what may be done in parallel and doing some additional stage redefinition. So far we have investigated parallel operation only in the context of combining stages that were not logically ordered. An extension of the idea of parallel operation involves the principle of *legitimate irrelevance*. This principle embodies the idea that it is worthwhile to perform a function, even if it is not necessary, if it requires no significant increment in time or circuitry. There is a related principle of *legitimate redundancy* that permits the same thing to be done a number of times if the repetition of the activity is overlapped with other things so that no additional sequential time is required because of the repetition. In effect, the principles say that one need not wait to find out if something needs to be done, that one can do it and then ignore it, and incur no penalty if it doesn't slow you down.

Let us apply this to the address formation phase. It looks as if the formation of an address is dependent on decode because decode must determine whether address formation is a requirement.

However, if the address formation mechanism is present and always available for use, what is the effect of doing the address formation all of the time, without waiting for decode to determine if it is necessary? We can perform the address calculation in parallel with the decode. Assuming that our earlier improvements are in place, that is, that we have a 1-cycle decode; are transferring IN, BR, and the address portion of IR in parallel; and have a 1-cycle address adder, the Gantt Chart (see Figure 9) pictures the execution of the LOAD instruction. The instruction executes in 17 elapsed machine cycles, 10 of which are attributable to memory delays.

In Figure 9 the registers of the processor are indicated on the Y axis, and time periods are indicated on the X axis. Line segments within a time interval indicate that the indicated register is busy during the time period. Numbers above line segments are keyed to the legend beneath the figure that explains what action is associated with each line segment. Time intervals during which there is no line segment associated with a register mean that the register is not being used in those intervals.

In addition to speeding the LOAD instruction, the parallel decode and address formation organization has the effect of smoothing the variations in instruction execution times. A memory referencing instruction may be fewer

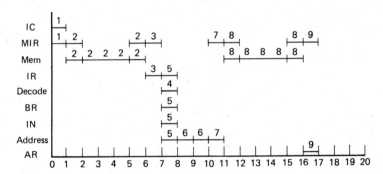

1. (IC) → MIR: request instruction.
2. MIR/MEMORY: address to memory; instruction to MIR.
3. (MIR) → IR: instruction to IR.
4. DECODE.
5. (IN), (BR), (IRadd) → address adder. Address elements to adder.
6. FORM ADDRESS OF OPERAND.
7. Address → MIR.
8. Operand → MIR.
9. (MIR) → AR.

**Figure 9.**   Gantt chart of improved LOAD.

cycles longer than a non-memory reference instruction. Notice also that the design of a 1-cycle decode becomes less critical. For memory referencing instructions, the 1-cycle decode does not speed performance at all. Whether it is then worthwhile depends on the architecture and the compiler and the relative frequency of use of instructions that do not reference memory.

The earlier suggestion of sending BR, IN, and IR to the adder in parallel required some extension to the adder to provide a holding place for the three inputs. The idea of doing address formation in parallel with decode is based on the assumption that all of the circuits required for address formation are always available. To do parallel decode and address formation there must be no circuits that are used in common by both functions. Sharing circuits introduces delays and constrains the amount of function performed in parallel. Any delays due to synchronization of circuit use would extend the time for both functions. The address adder and the decode circuitry must become effectively two independent units.

There is a coordinating relationship between decode and address formation in that decode, if it determines that a memory reference is not to be made with an instruction, must suppress the movement of the output of the address adder from being moved to the memory interface where it would cause a memory reference.

The reformed stage and cycle list for the example LOAD is now:

| | Stages | Cycles | Elapsed | Comments |
|---|---|---|---|---|
| 1. | (IC) to MIR | 1 | 1 | Request instruction. |
| 2. | (MEM) to MIR | 5 | 6 | Instruction to processor. |
| 3. | (MIR) to IR | 1 | 7 | To instruction register. |
| 4. | DECODE | 1 | 8 | Parallel decode. |
| 5. | IN, BR, IR to ADDER | 1 | 8 | Move address elements. |
| 6. | ADD IR, BR | 1 | 9 | Begin address formation. |
| 7. | ADD IN | 1 | 10 | Complete address formation. |
| 8. | (ADDOUT) to MIR | 1 | 11 | Request operand. |
| 9. | (MEM) to MIR | 5 | 16 | Operand to processor. |
| 10. | (MIR) to AR | 1 | 17 | Instruction complete. |

We have undertaken to increase the speed of the machine in two basic ways. We have made some functions faster, for example, the decode and the address formation addition, and we have redefined stages to increase parallel execution of functions within an instruction. By doing these things within the processor we have reduced the cycle count from 23 to 17 without any organization to address the memory speed imbalance problem.

We have also seen that there might be some rather subtle consideration necessary to determine whether to do the 1-cycle decode. A very high-performance model of the architecture might be able to afford to do both, but smaller models might have price/performance constraints that make it impossible to afford both the 1-cycle decode and the overlapped address formation. In general, choices between increasing the speed of a function and increasing the ability to do functions in parallel involve trade-offs and judgment. One must decide whether to use circuitry better to increase the speed of a function or to try to do more in parallel.

In more general terms, there is frequently a trade-off to be made between trying to speed the execution of all instructions and trying to speed the execution of particular instructions. There are price/performance constraints on organization that frequently force one to choose.

However, although trade-offs must be made, it is also true that function speed and overlap are frequently complementary. A function may be speeded up to assure that it is completely overlapped with another function rather than extended in time beyond it.

### 6. CONCLUDING REMARKS

The cost to the organization of reducing the cycle count from 23 to 17 machine cycles for the execution of this LOAD instruction would actually be quite modest. In general the cost of speeding very slow processor organizations is low relative to the increase in speed. However, there is a point reached when organizing for very fast machines where the increase in speed may be low proportional to the increase in the design, component, and manufacturing cost of a machine.

The organizational enhancements we undertook were limited by the simplicity of the architecture and the primitive nature of the instruction. In a machine with a large population of registers and more complex addressing conventions, much more might be done to speed the execution of instructions. The nature of an architecture somewhat limits the range of different organizations and the consequent range of price/performance ratios available to the architecture using a given technology.

To increase the speed of a model of a simple architecture beyond what we have discussed might be quite costly and represent serious increments to total system cost. The additional expense occurs because specialized organization and hardware are required for functions. Thus circuits that might be shared between functions must be duplicated. There is an expense associated with the design of specialized units and cost for manufacturing specialized components which may add significantly to the cost of the machine.

### QUESTIONS

1. What is I-time? E-time?

2. Why is it important to do address formation quickly?

3. What is the processor/memory speed imbalance problem?

4. What is the principle of legitimate irrelevance?

5. What is an instruction stage?

6. Why might the organization of a particular model of an architecture have to choose between overlapping address formation and fast decode?

# Chapter 16

# Extended Lookahead

## 1. INTER-INSTRUCTION OVERLAP

The discussion in the last chapter was limited to approaches for speeding a processor by making an instruction faster. Another way of speeding a processor is to design in such a way that there is some overlap between the execution of successor instructions.

Another glance at Figure 9 reveals that, during the total time of execution of the LOAD instruction, the registers of the system are very sparsely used. The IC is free for 16 of 17 cycles, the MIR is free for 9, the BR is not used for 16 of 17 cycles, and so on. Even the memory is free for 6 cycles. What this suggests is that there is an opportunity to undertake some additional work to utilize these unused facilities and consequently speed the machine. The method for achieving greater speed is to increase the parallel capability of the organization so that some processing of more than one instruction at a time can be undertaken.

An organization that undertakes to do I-time functions on an instruction while a predecessor is in E-time is called a *lookahead organization* or a *lookahead machine*. We suggested the idea of lookahead in Chapter 13 when discussing how far forward in an instruction stream a processor might have to look to find instructions that can be executed in parallel on a LOAD/STORE, as opposed to a three-address, architecture. Organizations that will not only perform I-time functions on one instruction while executing another but will allow the execution of more than one instruction at the same time are called *parallel organizations*.

This chapter introduces basic lookahead features into the organization. We are going to see what is involved so that operations on more than one instruction at a time may be undertaken. We limit ourselves to I-time and E-time overlap, however, and begin with a one-instruction lookahead. The goal is to have a successor instruction using the underutilized registers during periods when a predecessor is not using them.

The basic lookahead function is to move the instruction counter to point to a successor instruction while the processing of an instruction is occurring. It is almost universal for the address of the next instruction to be formed during the

I-time of a current instruction. Extensions to lookahead involve the actual acquisition of the successor instruction and the performance of functions up to the decode stage.

To increase the lookahead capability of the simple machine we have been discussing we are going to make two additional changes. First, we must reorganize the black box so that it has enough independent circuits so that more instruction stages may be performed in parallel with other stages. We have already noted that address formation and decode circuits must not be shared if address formation is to proceed in parallel with decode within an instruction. We must now increase circuit count and independence to assure further minimization of circuit sharing.

We are also going to introduce a new register, the *memory data register* (*MDR*) to hold the information that is being sent from memory. Henceforth the MIR will only be used for memory requests and the MDR will be used for the return of requested contents of memory.

Figure 10 is a picture of how we wish the processor to behave. Like Figure 9, Figure 10 is a Gantt Chart. It represents the time sequence of the activity of registers associated with the processing of the LOAD instruction and the preparation for the execution of a successor ADD instruction. The line segments represent the time periods during which registers are busy; their numbers are keyed to the legend. When two registers are communicating with each other during an interval, the same line segment number appears over the communicating registers.

The initial state of Figure 10 finds the address of the LOAD instruction already in the MIR. Before $T_0$, the IC has moved that address to MIR. During $T_0$ and $T_1$, the MIR is sending the request to the memory and the IC is being increased to point to the next instruction. By $T_1$ the address of the ADD instruction is formed in the IC.

The MIR is free at $T_1$. One difference between Figures 9 and 10 lies in the representation of the activity of the MIR as being distinct from the activity of the memory. Without the MDR no meaningful use could be made of the MIR until the memory had responded to the request. With the MDR as the recipient of the contents of memory, however, it is meaningful to represent MIR independently and to see when it is free, regardless of when memory is free. The MIR can receive addresses during certain cycles when the memory is busy.

Notice that we now also have a finer understanding of the basic machine cycles involved in a memory reference. We have been charging 5 basic machine cycles for a memory reference. In earlier chapters that time represented the time from the arrival of an address in MIR to the delivery of a value in MIR. Now Figure 10 shows that one of those cycles involves MIR/memory interaction to pass the address to the memory. It also shows that one of the cycles involves interaction between memory and MDR to place a value in MDR. Three cycles are used by the memory to actually find a value and prepare it for movement to the processor.

Since 1 cycle is necessary for the MIR to communicate with memory to pass on a request, the MIR is free to receive another request at $T_1$. In the interval

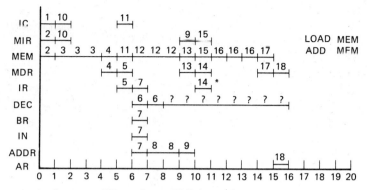

1. Increment IC to point to ADD instruction.
2. MIR gives address of LOAD to memory.
3. Memory fetches LOAD instruction.
4. Memory places LOAD instruction in MDR.
5. (MDR) → IR. LOAD instruction to IR.
6. DECODE LOAD.
7. Address elements from IRadd, IN, BR to address adder.
8. LOAD operand address formed.
9. LOAD operand address to MIR.
10. (IC) → MDR. Address of ADD to MIR.
11. MIR gives address of ADD to memory.
12. Memory fetches ADD instruction.
13. Memory places ADD instruction in MDR.
14. (MDR) → IR. ADD instruction to IR. HOLD*
15. MIR gives LOAD operand address to memory.
16. Memory fetches LOAD operand.
17. Memory places LOAD operand in MDR.
18. (MDR) → AR. LOAD operand to AR. Load Complete.

**Figure 10.**   Basic lookahead.

between $T_1$ and $T_2$, the IC places the address of the ADD instruction in MIR. At this time, however, we experience a memory delay. The memory is busy processing the request for the LOAD instruction. Until $T_5$ no reference to memory can be made. The machine effectively idles until the LOAD instruction arrives.

At $T_5$ the memory has finished transferring the LOAD instruction to the MDR. At this time it can accept the request for the ADD instruction that has been sitting in the MIR for 3 cycles. The following things now happen in parallel in the interval from $T_5$ to $T_6$:

1. The MIR communicates with the memory to pass on the address of the ADD instruction.
2. The LOAD instruction is moved from the MDR to the IR.

Naturally the ability for these two events to occur in parallel depends on there being sufficient circuits so that an MIR/memory interaction and an MDR/IR interaction do not interfere with each other. We must have completely distinct data paths and independent logic.

The delivery of the LOAD instruction to IR in the interval between $T_5$ and $T_6$ enables decode of the LOAD instruction to begin at $T_6$. The decode of Figure 10 is shown as a 2-cycle decode. Decode occurs during the interval $T_6$ to $T_8$. While decode is in progress, during the interval $T_6$ to $T_7$, the elements required to form an address are sent to the address adder. Thus the contents of the address portion of IR, of IN, and of BR arrive at the adder by the end of $T_7$. Figure 10 shows that an address will be formed during the interval from $T_7$ to $T_9$ and by the end of $T_{10}$ the address of an operand for the LOAD will have been placed in the MIR.

Notice the line segment on the decode line that extends from $T_8$ to $T_{16}$. We place a question mark over that line because it raises some questions about how the decode is actually accomplished and when decode circuits will become free to decode a successor instruction. $T_{16}$ is the point at which the LOAD instruction completes by virtue of the delivery of an operand to the AR. The actual interpretation of the instruction can be considered complete at $T_8$. That is, it is possible to set up all the signals necessary for the execution of a LOAD by that time. However, it may be necessary to retain certain information in the decode circuits throughout the entire execution of the instruction. If the design is done in that way no successor instruction can enter the decode stage until the completion of a predecessor. The limit of overlap is the entry to the decode stage. The furthest that lookahead can progress is to put a successor instruction into the IR and wait for the end of a predecessor E-time to free decode circuits.

A more aggressive lookahead design would define a place in the organization where whatever bits or signals necessary to perform execution would be held so that decode of the successor could begin the cycle after the successor had arrived in the IR. We look at motivations and mechanisms for this early successor decode later. With such a design the ? line segment would not appear on Figure 10. More significantly, the decode of the ADD would begin at $T_{11}$, and, by implication, the processing of a third instruction might begin then.

A less aggressive design might require that an instruction remain in the IR during all of its execution time. This would prohibit the movement of the ADD instruction from MDR to IR in the interval $T_{10}$ to $T_{11}$. That would have considerable effect on Figure 10 and the performance of the machine. For example, the ADD instruction would not move to IR until the interval $T_{16}$ to $T_{17}$, which would make the MDR unavailable to receive the LOAD operand in the interval $T_{14}$ to $T_{15}$ and unable to deliver the value to the AR by $T_{16}$. Thus the retention of the LOAD instruction in IR during its entire execution would not only delay the ADD instruction but actually extend the length of the LOAD. Therefore, it is necessary to free the IR to increase the speed of the machine. A further increase is achieved by freeing decode circuits as quickly as possible. Figure 10 shows that the lookahead is stopped with the delivery of the ADD to IR and that the decoding of ADD does not begin until the completion of the LOAD. This is based on the assumption that the decode functions are not then available.

The processor that will support the overlap of Figure 10 is shown as Figure 11. We see:

1.  The MDR with a path to the AR to allow delivery of operands to the AR and its associated extension register.
2.  The MDR with a path to the IR to allow delivery of instructions to IR.
3.  The MIR with a path to the IC to allow IC placement of instruction addresses in MIR.

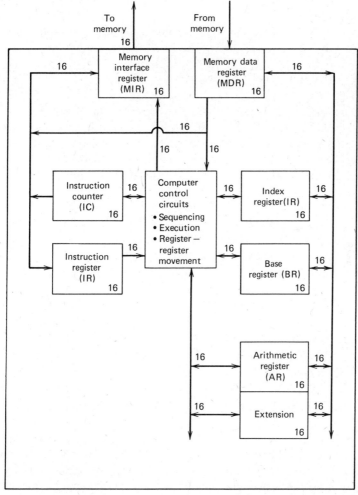

**Figure 11.**  Basic lookahead processor.

4.  A path from the control circuits black box to the MIR to allow operand addresses to move from the address adder to the MIR.

5.  The path between MDR and AR also connects the MDR to the BR and the IN. This provides for the placement of values in the IN and the BR when these values are specified as occurring in memory locations. Independent paths to AR, IN, and BR are not provided because the design does not anticipate the need for parallel movement between MDR and these registers.

6.  A path between IN, BR, and IR and the control box to allow input of address elements to the address adder and to provide for the placement of values in IN and BR when these values are specified as immediate.

7.  A path between the control box and IC to provide for the increment of IC and for the placement of addresses in IC as a result of branching operations.

8.  A path between IR and the control box to support decoding functions.


## 2.  STORING VALUES INTO MEMORY

The discussion so far has presented the organization of a processor in the context of the LOAD and ADD instructions. Each of these is characterized by the fact that a memory reference for an operand involves only a fetch from memory. We have not addressed the nature of storing an operand into memory. We undertake to do so now and show how an MIR only and an MIR and MDR design would behave.

The storing of a value in memory involves the provision of an address where the data is to be stored and data that is to be stored at that address. With only an MIR at the memory interface a store into memory will take longer than a fetch from memory because it is necessary to provide the MIR with the address and the data sequentially. A reasonable sequence for storing is as follows:

1.  Send the formed operand address in the STORE instruction to the MIR. This movement from the address adder to MIR takes 1 machine cycle.

2.  Pass the address from the MIR to memory. This requires 1 cycle and frees the MIR at the completion of the cycle.

3.  Send the data in the AR (or the extension) to the MIR. This requires 1 cycle.

4.  Send the data from the MIR to the memory. This also requires 1 cycle.

The addition of the MDR presents a number of possible faster designs. When we consider moving data from memory to the processor we use MIR as an address register and MDR as a data register. When storing data we may use the MIR and the MDR in the same way, but we change the timing of the placement of addresses in MIR and of data in MDR to permit us to overlap the loading of data into MDR with the transfer of the address to MIR. Alternatively we may overlap the transfer of data to MDR with an MIR/memory interaction where MIR passes the storage address to memory. The processor performs in the following manner:

1.  Send address to MIR (and send data to MDR?).
2.  Send address to memory (and send data to MDR?).
3.  Send data to memory.

If MIR and MDR are to be loaded in parallel it is necessary to have independent data paths between the AR from which the value is coming and the address adder from which the address is coming. The processor in Figure 11 has such independent paths.

If both data and address are sent in the first cycle, it is possible to design a memory interface where the contents of MIR and MDR are transferred to the memory in parallel. This design reduces the number of sequential cycles needed for a store of data on the processor side. If it is not possible for a memory to interact with MIR and MDR in parallel then there will be an MIR/memory interaction to pass the address, followed by an MDR/memory interaction to pass the value. Such a sequence is still faster than an MIR only sequence:

| MIR Only | Cycles | MDR/MIR No Parallel MDR/MIR/MEMORY | Cycles |
|---|---|---|---|
| ADDRESS to MIR | 1 | ADDRESS to MIR, VALUE to MDR | 1 |
| MIR/Memory | 1 | MIR/Memory | 1 |
| VALUE to MIR | 1 | MDR/Memory | 1 |
| MIR/Memory | 1 | Memory function | 3 |
| Memory function | 3 | | |
| Total | 7 | Total | 6 |

The parallel Memory/MDR/MIR interface eliminates 1 more cycle from the memory store operation. Thus the transfer of address and data to the memory is accomplished in 1 cycle, reducing to 5 cycles the time required for a store.

The reduction of cycles required for a store operation from 7 to 6 or 5 may not seem significant. However, it represents a very significant percentage reduction in elapsed time for a function that occurs very frequently and will significantly speed the machine.

The above discussion assumed a 1-word 16-bit architecture with 16-bit wide data paths. A design that uses shorter data paths or an architecture that is byte-oriented but has the concept of multibyte words might use multiple cycles to transfer data from registers and between the memory interface registers and the memory.

## 3.  MEMORY READ AND WRITE CONTENTION

The use of MDR as a space for values to be stored in memory is not without its penalty. In the overlap design of Figure 10 we relied on the ability to submit a new address for a memory fetch while a previous memory fetch was going on because the results would be placed in MDR. If, however, MDR is going to be used as the location to place data for a memory write operation, then a contention for MDR will develop between incoming and outgoing data. A STORE operation may be delayed until a unit of requested data is delivered to the processor, an operand delivery may be delayed until a STORE is completed.

In order to avoid interference between memory references in different directions we can add an additional register. We call this register the *memory storage register (MSR)*. Now the memory interface contains three registers, MIR for the address of stores and loads with memory, MDR for the return of data from memory, and MSR for the values to be stored in memory. With this design a request to bring a value from memory may be overlapped with a request to put a value in memory.

However, there is still a contention for the MIR. With a memory that cannot support any notion of parallel operation, the contention in MIR between store and load requests is not critical. However, anticipating later comment, we add a fourth memory interface register, the *memory address register (MAR)*. This register is used to place the addresses of memory locations in which values will be stored. The MIR is used only for load addresses. We now appropriately rename it the *memory fetch register (MFR)*. A full justification for this population of registers will wait until we discuss memory design. Figure 12 shows the organization. Figure 12 differs from Figure 11 only in the elaboration of the memory interface registers. Data is moved to memory through the use of the MAR and the MSR. Instructions and data come from memory through the use of the MFR and the MDR.

## 4.  MEMORY REFERENCE DELAYS

The overlap in Figure 10 shows a single instance of memory delay when the request for the ADD instruction was delayed because the memory was busy fetching the LOAD instruction.

In general, a population of instructions that is being overlapped in the manner of the last section frequently runs into memory reference delays. These delays occur between instruction fetches, between operand fetch and instruction fetch, between operand store and instruction fetch, and between operand store and operand fetch, depending on the specific pattern of instructions and the specific details of how much of one instruction is overlapped with another.

As a result, instructions may be delayed because of the overlap, and the full benefit of overlap may not be realized. In order to reduce the impact of memory reference delays due to contention and interference, it is necessary to elaborate

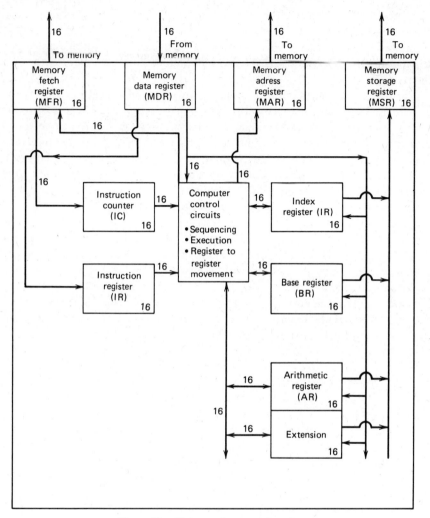

**Figure 12.** Elaborated memory interface.

the design of memory in order to make it possible that not all references to memory result in a contention.

It is now necessary to be sure that we make a distinction between memory contentions and the memory/processor speed imbalance represented by the fact that we are associating 3 basic machine cycles with each memory reference. The memory contention problem is related to the memory speed problem, but it must be understood in its own right. In the context of this chapter we may be disappointed in performance because the memory is busy and not directly because it is slow. There are techniques for addressing the busy problem that are independent of the speed problem.

The goal of a solution to the memory contention problem is to design in such a way that a reference to memory will not require more than the 3 cycles a memory operation requires. We encounter effective delays of more than 3 cycles when a memory interface register cannot communicate with memory because the memory is busy.

For the rest of this chapter we assume a memory that has the property that any interaction with memory will require only 3 cycles; in effect, instruction fetches, operand loads, and operand stores may proceed at the same time if no register contention occurs. This assumption is necessary if we are to investigate extended lookahead designs with an anticipation that they will be profitable. In order to do more effective lookahead we must have a memory that minimizes memory reference time, a memory that is never "busy" and upon which queues do not form. Successive instructions must be obtainable without chronic memory delays.

Memory design, as we see in Chapter 18, has a general goal of minimizing processor delays due to memory reference. Some of the memory design techniques attempt, as here, only to eliminate contention delays so that memory response time is always the memory time. Other techniques attempt to develop a memory so that the average time to respond to a processor request is actually less than fundamental memory speeds. The assumption of this chapter is simply that contention delays do not exist.

## 5. INCREASING LOOKAHEAD

Increasing the lookahead capability of a processor involves an increase in the number of instructions that can be processed through I-stages at the same time and an extension of the point at which an instruction is delayed from further processing. However, we stop short of allowing parallel execution in this chapter. Basic to extended lookahead is a memory that can provide instructions at a maximum rate. We also require some elaboration of the processor.

The reason for suspending progress with the ADD instruction in Figure 10 was that decode could not be undertaken while the previous LOAD instruction was still in the machine. Even though the LOAD was in E-time, decode circuits could not be used for the ADD.

In order to break this bottleneck, we increase and reorganize the circuits in the black box. We define a decode output area that enables the decoding of a successor instruction to proceed with the execution of a predecessor. This is possible because the decode cycle no longer changes the state of the circuits of the machine and interferes with the completion of a prior instruction. The output of the decode stage is merely placed in an output register. Use of this output occurs at the completion of the previous instructions.

We have been adding circuits to the control box to minimize sharing and enable increased parallel function. We now formalize our notions of the circuits in the control box around the notions of I-time, E-time, and decode by defining an element called an *I-unit*, or *I-box*, or *Instruction Processor*, or *Instruction*

*Pre-Processor Function (IPPF)* that performs all I-time functions and whose link with a completely self-contained *E-unit* or *E-box* is through a *decode output register (DOR)*. This decode output register contains a set of functional directive bits and the fully formed operand address required by the E-unit for fetching operands. The address adder output register becomes the address portion of the DOR and the operand request addresses go from the DOR to the register at the memory interface used to hold operand fetch request or store request addresses.

Although the I-unit and E-unit must communicate through a shared register, the E-unit shares no circuits with the I-unit. (I-unit and I-box, and E-unit and E-box are interchangeable synonyms.) We now have the basis for two forms of parallel function in the design. There is some overlap between I-time functions in the I-box, and there is overlap between I-time and E-time. The processor we have now defined is shown as Figure 13.

## 6. DETAILS OF I/E FUNCTION AND RELATIONSHIP

In Figure 13 the I-box has access to the IC, the DOR, the IR, the IN, the BR, and the set of memory interface registers (MFR, MDR, MAR, and MSR). It has no access to the AR or its extension. The I-box contains the function to update the IC, the address adder, and the decode circuits. The E-box has access to the DOR, the AR and its extension, the IC, the IN, the BR, and the memory interface registers. A discussion of these paths will help us understand the relationship between architecture and organization and the nature of organization more fully.

### I-Unit Relationships and Functions

1. Update IC to develop next instruction address.
2. Cause transfer of branch address from address adder to IC.
3. Cause transfer from IC to MFR to request instruction.
4. Cause transfer from MDR to IR of requested instruction.
5. Cause transfer from IN, IR, and BR to address adder.
6. Cause output of address adder to go to DOR.
7. Cause coded results of decode stages to go to DOR.

### E-Unit Relationships and Functions

1. Cause addresses in DOR to go to MAR to provide address for operand.
2. Cause operand to go from AR or extension to MSR to provide value for storing in STORE instructions.
3. Execute branch instructions and inform I-box to place branch address in IC.
4. Execute all instructions that affect values in AR or its extensions.
5. Execute instructions that affect values in IN and BR.

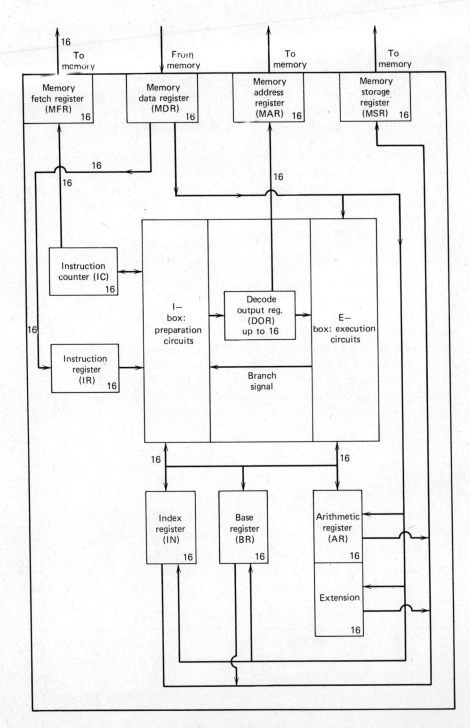

**Figure 13.** I-E separate units.

208

The above functional split is really just a starting point for discussion. A large number of variations in functional splits between I and E-units are possible. For example, it might be possible to impose the formation of addresses on the E-unit. It is possible to place conditional branch determination in the I-unit, effectively turning it into an E-unit for this class of instruction.

In general, the flexibility in determining the split of work between an I-box and an E-box is based on an appreciation of where bottlenecks will occur. If the E-box is chronically slower than the I-box, that is, the execution cycles of most instructions are more than the instruction preparation cycles, then it might be worthwhile to relieve the E-box of some work and have some instructions executed by I. A danger in this, however, is that the I-box may be delayed in its lookahead while it resolves the branch condition or performs additional logic of other kinds.

Two basic approaches to balancing I and E so that bottlenecks do not occur are to move more functions common to all or most instructions to the I-box or to have the I-box undertake to execute some class of instructions. We saw above that conditional branch execution might be such a class. The processor of Figure 13 could not do this in an architecture where a comparand came out of the AR. We would have to define a path to AR or develop some condition code setting logic for the I-box to inspect the status of the contents of the AR.

Another potential function that can be moved to I-time is the actual request of an operand. If address formation is performed by an address adder in the I-box it is possible to have the I-box move the address to MFR and not place the formed address in the DOR. Thus requests for operands would already be made by the time the E-unit accessed the DOR. The E-unit would then merely wait until it received a signal that the operand had arrived. This would effectively move the execution of all loading instructions to the I-box. It would require, of course, that the I-box have access to the arithmetic registers. In designs where the I-box function is so extended, the E-box performs only the arithmetic, boolean, and shift instructions (the functional instructions), and the I-box executes all nonfunctional instructions.

In Figure 13 the sharing of registers between E and I-boxes suggests that there may be times when, though they share no logic, the two units may get into each other's way because they need access to the shared registers for various independent functions.

Naturally the specific design must respond not only to goals for achieving parallel operation but to the details of the architecture. The population of registers, the specific instruction set, and the addressing techniques available all define what paths are necessary to perform the functions, what paths are necessary to perform certain functions in parallel, and what circuits are necessary to perform functions in parallel. The determination of what unique and unshared circuits and paths are appropriate depends on an analysis of how frequently there may be contention for the same path under certain assumptions of instruction execution patterns.

It is necessary for us to be very aware that as we increase the overlap potential of a design we are increasing the difficulty of predicting just how a processor will perform. Sometimes we will experience a memory delay in sequences of instructions, sometimes we will not. Sometimes we will run into sequences of instructions that do not reference memory because they are working in registers or because they are immediate instructions. Sometimes, depending on the length of time an instruction stays in the E-box, we may delay certain I-time functions. We face a problem in that, although the I-time for an instruction can be held constant or relatively constant, the E-time for a set of instructions can vary significantly if the instruction set contains a rich set of arithmetics including various forms of multiply and divide. More elaborate designs that continue from the design we have so far described, primarily address ways of breaking bottlenecks between semi-autonomous asynchronous units in order to maximize their utilization.

### 7. BUFFERING AND EXTENDED LOOKAHEAD

In the face of instruction streams of various patterns we wish to maximize the utilization of the I-box and the E-box in order to get maximum benefit from the proliferation of circuits that make them autonomous units. The goal of the design must be to provide units with work whenever they are free and to minimize the incidence of conditions that delay the completion of their work.

In the presence of semi-autonomous units with different rates of completion, a common design approach is to provide a buffering mechanism between the units. The buffer provides a place for the producing unit to put its output and continue to new work. It also defines a repository of work for a consuming unit. The presence of a buffer allows a producing I-box and a consuming E-box to operate in parallel a higher percentage of the time than they could if they were more closely coupled.

The relationship between I-box and E-box in the Figure 13 design has a number of potential problems. The I-box may not be able to provide a steady flow of instructions to the E-box because the I-box must refer to memory for every instruction and experience the processor/memory speed imbalance. In addition, although we have been assuming that there will be no delays in the system because the memory is busy, the available techniques for avoiding these delays due to contention for memory reference cannot really guarantee there will be no delay. They can only attempt to minimize its occurrence. Therefore there will be intermittent delays at the memory interface that will cause I or E-units to fall behind each other.

Another potential problem is that the decoding of instructions will be delayed and lookahead will falter because the DOR (decode output register) is not free. An instruction with a long execution time will prevent the I-box from putting an instruction in the DOR for long enough to prevent the I-box from continuing lookahead.

We address these problems by turning the IR and the DOR from a single register into a buffered family of registers, one set on each side of the I-unit. These buffers will be associated with a pointer mechanism that indicates in a round robin fashion which is the currently free register (for the IR buffer) and which are the currently free and the currently used registers for the DOR buffer.

As an example, if we place a set of four IR buffer registers that can hold instructions behind the IR register, the processor will undertake to update IC and request instructions whenever one of these registers is free. The intent is to try to provide a repository of instructions to the I-box that are accessible at processor and not at memory speeds and consequently to increase the rate at which I-functions can be applied to instructions. This will increase the probability that whenever the E-box is free there will be an instruction prepared for it, ready to be executed. Ideally the I-box may process multiple instructions through I-phases and get rather far ahead of the E-boxes.

The DOR register set on the other side of the I-box provides a place where instructions can be held after they have gone through a decoding phase. This buffer serves two purposes. It provides a place for the I-box to put instructions when they are ready for execution so that it can free locations in the IR buffer on its other side and maximize the rate of instruction request. It also increases the probability that the E-box will find an instruction ready for execution whenever it is free. Instructions are sent to the E-box from the DOR buffer area at a rate determined by the speed of the E-box execution of instructions.

We now have a system in which there are constant requests to memory for instructions (whenever an instruction buffer register to the left of the I-box is free), providing for more intense utilization of the preliminary decode and address formation circuits. We also have a system in which there is a collected pool of instructions for the E-unit so that the E-unit can keep ahead of the I-unit and not be delayed by the I-unit if it runs into delays at the memory interface.

We can similarly elaborate on the memory interface and turn the MFR, MDR, MAR, and MSR into buffers. We may, of course, buffer to any depth, the selection of depth requires a good deal of analysis and simulation. The principle, however, can be demonstrated with just a backup buffer behind each of the family of memory interface registers. Fetch addresses may now be queued and no delay will occur because of a busy MFR.

On the store side, where values are being sent from an execution unit to memory, values can be drained from the arithmetic register and successor instructions undertaken even if the memory is busy.

We are now close to the limit of increased execution with an overlap machine. The machine we have designed looks like Figure 14. We now look into techniques that go beyond lookahead to permit more than one instruction at a time to be in execution stages.

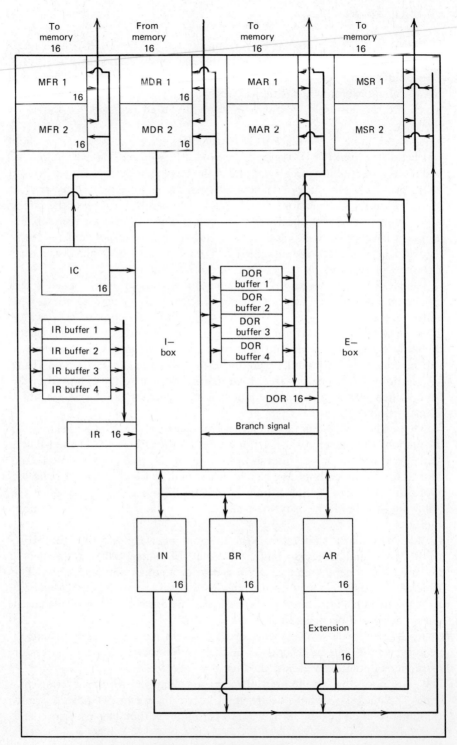

**Figure 14.** Buffered machine.

## QUESTIONS

1.  What is lookahead?

2.  Why did the number of registers at the memory interface grow from one to four?

3.  What are the basic functions of an I-unit?

4.  What is the intention of the register called the DOR?

5.  How does the processor in Figure 14 change if the I-unit is to execute conditional branch instructions?

6.  What is the purpose of buffering instructions to the I-box?

7.  What is the purpose of buffering the output of the I-box?

8.  Draw a Gantt Chart of the operation of the processor of Figure 14 for the instruction sequence:

    > LOAD R1, A
    > LOAD R2, B
    > ADD R1, R2
    > STORE R1
    > MPY R1, C
    > SUB R1, D
    > STORE R1

# Chapter 17

# Parallel Instruction Execution

### 1. MULTIPLE INSTRUCTION EXECUTION

By the nature of lookahead design we permit only as much parallel operation as is possible between instructions independent of any potential logical relationship with each other. The goal of lookahead is to overlap instruction preparation with execution and to effectively reduce the time duration for the execution of an instruction to those cycles required for execution. This goal is achieved when all I-time functions are overlapped with E-time functions for predecessor instructions.

The next step in speeding the instruction processing ability of a processor is to undertake to execute instructions so that more than one of them may actually run to completion during the same time intervals. In order for two or more instructions to be fully retired in the same time period, or to enter their execution stages in the same time period, two things are necessary:

1.  The physical circuits must exist for the parallel execution of the basic machine cycles associated with execution stages. These circuits include non-shared control logic, non-shared register space, and non-shared data paths between storage elements. The duplication of circuits associated with lookahead is extended to the organization of E-units.
2.  The machine has the ability to recognize logical relations between instructions in order to determine if one instruction is dependent on the prior execution of another.

In order to have parallel instruction execution it is necessary to have both sufficient circuits and logical independence between instructions. Machines that allow for such parallel instruction execution are called *parallel* and *pipelined machines*. Parallel may mean merely that there is a parallel E-time capability. The word may also be used so that it implies that there are separate instruction execution units for various classes of instruction. Pipeline means that the execution stages of an instruction are broken down so that multiple instructions can be handled by the same execution unit, performing different stages on different instructions at the same time.

## 2. CONSIDERATIONS IN MULTIPLE E-BOX DESIGN

Figure 15 is a partial representation of a processor with multiple execution units. There is an I-unit and a set of E-units. Some designs call for an ability for any E-unit to be able to execute any instruction. These are *homogeneous E-box designs*. More common in contemporary high-performance uniprocessor design

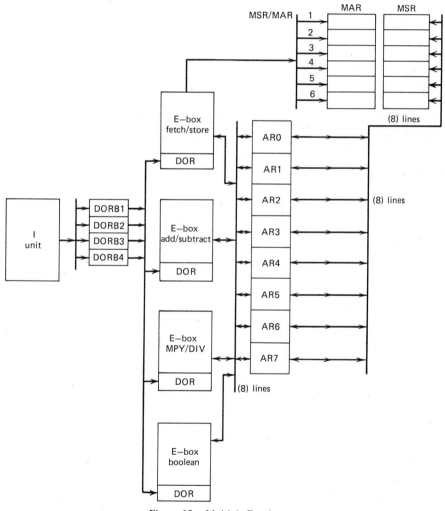

**Figure 15.** Multiple E-units.

is for various E-boxes to be dedicated to specific instructions and designed for particularly efficient execution of the class of instruction it executes. In this way speed is achieved not only by overlapping instruction executions but by unique design for effective execution within each E-unit. Such designs are *heterogeneous E-box designs*. Figure 15 indicates a heterogeneous design with a population of E-boxes. There is a fetch/store unit that handles all references to memory, an add/subtract unit, a multiply/divide unit, and a boolean unit shown. The population of E-units in Figure 15 is illustrative and not complete.

In this figure each E-box has common access to the DOR buffer area and brings an instruction into its own private DOR when it is ready to execute a new instruction. Each E-box has access to a set of arithmetic registers which may also be used as index registers. The arithmetic registers have common access to a set of buffered memory registers for storage of operands.

There are a large number of design considerations that are relevant to the design of processors that have populations of E-boxes. One must consider:

1. The population of E-boxes and how functions are distributed.
2. The relationship between the I-box and the E-boxes as regards the delivery and allocation of instructions.
3. The delivery of operands to E-boxes.
4. The relationship between E-boxes and the operand registers.
5. Synchronization between E-boxes.
6. Internal buffering and design within an E-box.


3.  POPULATIONS OF E-BOXES

It is necessary to determine how many E-boxes there will be and how work is to be divided between them. The E-box population must be defined in terms of some notion of the influence of the architecture on the specific sequence of instructions that will occur. The desire for fast execution of certain slow instructions will justify separate E-boxes for those instructions. For parallel operation as many E-boxes as will effectively perform parallel instruction execution are desirable. Thus one must understand programming flow patterns to determine how instructions that may be logically unrelated and executable in parallel may occur. This parallel potential may be achieved by a compiler that rearranges instructions to maximize parallel potential. The rearrangement of instructions by the compiler increases the probability that the lookahead mechanism, as it searches forward on an instruction list, will find instructions that can be executed in parallel because they have no logical relationship with each other. We have seen that the probability of finding unrelated instructions depends not only on the depth of the lookahead but also on the architecture. The architecture determines the extent to which a particular function on data is represented by one instruction. Later we see that there is some dynamic instruction resequencing potential that may be placed in hardware design to augment compiler arrangement and increase the probability of finding parallel instructions.

The ultimate E-box design is to have an E-box for each individual instruction across function and data representation. Such a design would not be profitable since instruction interdependence and any reasonable lookahead mechanism could rarely find an instruction for each execution unit at all times. Therefore E-boxes are commonly grouped around some organizing principle. Two common groupings follow:

1. *Grouping by Data Representation.* In such an organization E-boxes are allocated function as binary units, decimal units, or floating point units. All arithmetic operations applied to data of a particular form are executed by the unit designed to handle that form.
2. *Grouping by Generic Function.* In this organization each E-box is designated by function (adder, multiplier, divider, etc.) and performs that function on all data regardless of representation.

There are many possible and actual variations and combinations of these two approaches that may come from either price/performance judgments or the architecture. If, for example, the architecture has only a floating point divide instruction, then a divide E-box will only handle data in that form.

## 4. DELIVERY OF INSTRUCTIONS TO E-BOXES

Figure 15 shows a processor with a DOR buffer area between the I-box and a set of E-boxes. Our last elaboration of the lookahead machine defined the DOR buffer in order to smooth the potentially bottlenecking relationships between I and E due to variations in E-time and to potential memory delay problems. Such an intervening buffer may well exist on a multiple E-box machine. It serves the same general purpose as it does on the lookahead machine. If such a buffer does exist, the I-unit will fill the DOR buffer with instructions at a rate partially independent of the rate of instruction execution. Figure 15 shows that each E-box has its own DOR in addition to the buffer. Although the figure shows a heterogeneous E-box design, the concept may apply to homogeneous designs. It is desirable to remove instructions from the DOR buffer at the fastest possible rate in order to free DOR buffer space. In a homogeneous system, each E-box takes the next instruction from the next DOR buffer register as it completes an instruction and becomes available.

In a heterogeneous design each E-box may inspect a code placed by the I-box in the DOR word format to determine whether an instruction is of the class that it performs. If it is, the E-box transfers the instruction to its own DOR. In some designs the DOR buffer area is not used as a general instruction buffer for any class of instruction. Instead each E-box has its own DOR buffer area. The I-box sends instructions to appropriate E-boxes on the basis of function and availability. Thus each E-box has a private collection of instructions within its own DOR buffer area. Of course it is not necessary for an E-box to have buffered DORs even if there is no general DOR buffer area. Whether to buffer E-boxes depends on the impact of the I-box encountering situations where it cannot pass an instruction to an E-box because the E-box DOR holds an instruction that is currently being executed. These delays mean that the I-box may be unable to continue lookahead because it cannot free its decode circuits. The inability to free its decode circuits may result in an inability to process instructions in its IR buffer area, and a consequent inability to continue bringing instructions to IR locations causing a suspension of lookahead. This suspension of lookahead is serious on a parallel machine because it means that any number of E-boxes may be denied instructions due to an interlock between the I-box and a single E-box.

### 5. DELIVERY OF OPERANDS TO E-BOXES

E-boxes may be made responsible for acquiring their own operands. The operand address is placed in the DOR by the I-box and the E-box requests the operand from memory.

Alternatively an *operand forwarding* design may be used whereby the I-box makes the request for an operand at the time that it forms the full address. There are some architectural considerations that apply.

A register file architecture that does not allow memory addressing in functional instructions must execute LOAD or PUSH operations before functional instructions. These LOAD or PUSH operations may be executed by the I-box, as we saw in the previous chapter, or they may be executed by E-boxes that have as their special function the execution of memory references. Since explicit LOAD operations are going to be executed, each of these LOADs will specify an operand register in which to place the operand. An E-box executes when it has received an instruction and when it determines that the operands have arrived in the arithmetic registers.

Accumulator model architectures that permit operand addressing in arithmetic instructions may accomplish operand forwarding with some additional design complexity. In effect the arithmetic instruction that contains a memory reference may be decomposed into two separate instructions.

| Instructions: | LOAD R1, A |
| | ADD R1, B |
| Effective Instruction: | LOAD R1, A |
| | LOAD R?, B |
| | ADD R1, R? |

The R? suggests a problem with instruction transformation of this type. It is difficult for these machines to predict the contents of arithmetic registers at the time they are requesting memory fetches. Since registers are program addressable and the compiler has done register assignments in the generated code, the hardware is not free to dynamically redefine register use. Therefore the ? suggests that operands must be forwarded to a place in the organization that has not yet been defined. This place may be an internal operand register that exists within the E-box. An E-box, therefore, in addition to having a DOR or a DOR buffer of its own, may also have a private place in which to hold operands that are being forwarded by an I-box that cannot use an arithmetic register. This area may itself be buffered so that an E-box may accumulate a set of operands that it may associate with its collection of DOR instructions. The pattern of execution, therefore, is that an I-box forms an address and designates an internal E-box operand register as the place for memory to send the operand. It requests the forwarding of the operand to the E-box register and releases the instruction to the E-box. The E-box, on encountering the instruction, checks that referenced operands have arrived in the general registers and in its own internal operand buffer.

In designs where a number of requests may be given to memory and be outstanding at the same time, it is necessary to preserve the E-box register destination address so that the final location of data delivered by the memory will be known. A simple method for doing this is to pass the destination to the memory and have the memory return the destination address when it returns the data.

## 6. RELATIONSHIP BETWEEN E-BOXES AND REGISTERS

If E-boxes are homogeneous, they will require common access to the operand register set and any register must be available to any E-box since any E-box may execute an instruction referring to any register. If the E-boxes are heterogeneous then the association of E-boxes with the register population depends on the specifics of the register model and the specifics of E-box function.

We have mentioned before that E-box populations may be organized in various ways by function and data representation. If the register population has specialized operand registers, such as floating point and binary, then it is natural to split E-boxes by data representation on a high-performance model of the architecture. Floating point E-boxes will have access to the floating point registers, binary boxes to the binary registers, and so on.

If there are no specialized registers then all the E-boxes must access the same register set regardless of the organizing principle of the E-boxes. Thus if all arithmetic functions are done in the same set of general-purpose operand registers, then all of these registers must be available to all E-units.

It is also desirable that each E-box has independent access to any register so that it may retrieve and store values in any register associated with an instruction it is executing without encountering delays in access to the register set.

Figure 15 shows an E-box population in which each E-box has independent access to all operand registers. The figure also indicates that each arithmetic register has an independent path to a set of memory storage registers. This provides the ability for values to be moved from different ARs to the memory interface in parallel. To support the parallel movement of data from ARs, both the memory interface register holding data values to be stored and the register holding storage addresses are buffered sets. Thus a queue of requests for storage is maintained at MAR and at MSR. This enables registers participating in computation to have values removed and become available again at a rate that is unrelated to the speed of memory. The fast availability of registers is a fundamental performance requirement for a high-performance parallel organization.

The architecture implied by Figure 15 is a register file in which only LOAD and STORE instructions reference memory. Thus a STORE instruction transfers the address from the fetch/store E-unit and the value from a referenced register to the MAR/MSR buffers. It is for this reason that only the fetch/store E-unit has paths to the MAR. A similar scheme, not shown on Figure 15, applies to memory registers that request data from memory.

## 7.  INSTRUCTION SEQUENCING

We now address problems that must come to mind when considering the execution of instructions in parallel.

1.  How does one guarantee that instructions will be executed in proper sequence?
2.  How does one identify sequences of instructions that can be executed together?

One wishes to avoid the discouraging results of the following instruction sequence on a machine with a memory fetch unit, a memory store unit, and independent add and multiply E-units.

$$D=(A+B)*C:$$

1.  LOAD R1, A
2.  ADD R1, B
3.  MPY R1, C
4.  STORE R1, D

Figure 16 shows what might happen if we do not address the sequencing issue. We assume independent E-units for each instruction and the timing assumptions shown on the Figure. The ADD instruction begins before the LOAD completes. The MPY instruction begins before the completion of the ADD. But this is not too serious since the STORE begins just in time to store A into D.

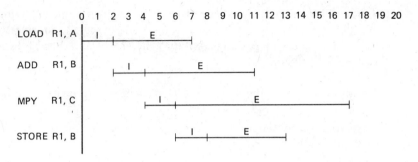

Assume:

1. Independent I, LOAD, ADD, MPY, STORE units.
2. 2-cycle I-time.
3. 5-cycle E-time for LOAD and STORE.
4. 7-cycle E-time for ADD.
5. 11-cycle E-time for MPY.

**Figure 16.**  Unsynchronized instructions.

The problem is that the code is entirely sequential, each instruction is dependent on a predecessor and each instruction has an input that is the output of a previous instruction. We need some mechanism for assuring that the ADD does not begin until the LOAD is complete, and that the MULTIPLY does not begin until the ADD is complete.

## 7.1.  Sequencing Techniques

In order to recognize and perform instances of parallel execution the following are necessary:

1. The processor must keep a running representation of the use status of its registers.
2. The processor must keep a running representation of the reference pattern for memory locations.

The first of these is relatively simpler than the second. A register may be defined to have three basic states: *unused*, *source*, and *sink*. An unused register is not involved in any instruction currently in the execution units. A source register is one that will provide an operand for an instruction in the execution units. Thus, in STORE R1, B, the reference to R1 puts it in source condition. A sink register is one that is going to receive an operand requested by an instruction from memory or from another register. Thus, in LOAD R1, A, the reference to R1, puts it in sink condition.

Given a set of instructions collected by lookahead and placed in an IR buffer mechanism, an I-unit can determine the relationships between instructions by the intersection of their register references. It may then impose sequencing rules to insure that instructions dependent on each other are executed sequentially. Details will differ from design to design, but the general approach is the following:

1.  Maintain a record of the status of all registers that indicates whether registers are unused, source, or sink.

2.  If an instruction contains a register reference, determine the status of the referenced register.

3.  If the referenced register is unused, mark the register as source or sink, depending on the nature of the reference, and forward the instruction to a DOR area.

4.  If the referenced register is source or sink hold the instruction in the I-box until a signal is received that indicates that the operand has arrived in a sink register or has been taken from a source register.

The above simple scheme will make it impossible for E-units to execute instructions out of order because instructions will not be released to the E-units when a logical predecessor is being executed. Thus the instruction sequence shown above will behave as follows:

1.  LOAD R1, A: Place R1 in sink status until A arrives. Next instruction will not be released until A arrives.

2.  ADD R1, B: Place R1 in sink status until sum of A+B arrives. R1 is used as source and sink but sink is overriding. Next instruction not released until ADD is complete.

3.  MPY R1, C: Place R1 in sink status. Hold STORE.

4.  STORE R1, D: Place R1 in source status.

The above simple sequencing scheme will prevent the difficulties of unsynchronized multiple E-boxes. It has the negative effect, however, of holding instructions in the I-unit until a predecessor is complete. This may reduce the depth of lookahead by extending the time that it takes for the I-box to release an instruction. Thus the possibility of finding unrelated instructions is reduced. More elaborate synchronizing mechanisms undertake to enable the I-box to pass on an instruction to a DOR area, either a general buffered DOR or DORs in E-units.

When it determines a logical relationship between instructions by virtue of intersecting register references, the I-box may forward the instruction to a DOR or E-unit with a flag indicating that referenced registers are to be checked for sink or source status before execution. Register status checking is moved from the I-box to E-boxes that become responsible for coordination. This allows the I-box to continue lookahead, searching for independent instructions and experiencing minimal delay due to instruction interdependence. One use of DOR buffers in an E-box is to maximize the number of instructions E-boxes may hold to help sustain the lookahead rate by providing maximum probability that the I-box will find an area in which to place an instruction.

It is also possible to make a distinction between a source and a sink reference. A source reference by a previous instruction may complete earlier than a sink reference. Consequently, some machines forward source references to an E-unit with a signal to check that the source reference is complete but hold onto sink references.

We are going to investigate the behavior of a machine with multiple execution units, applying the sequencing rules of the type we have just proposed. In doing this we make the assumption that our "magic memory," capable of parallel fetch and store, is operative. We also assume that IR instruction buffering around the I-box is sufficient to guarantee that we have the instructions on the list always available in the IR buffer. A summary of the assumptions of the example follow:

1. The I-box will always find a successor instruction in the IR buffer area. In this example we discount IC functions and do not represent activity involved in filling IR buffers.

2. E-units have buffers for DOR and do not cause delays due to there being no place to forward an instruction.

3. A reference to memory may be overlapped with another reference to memory. In effect the memory can deliver and store in parallel up to a limit of two requests.

4. There are no references to memory for arithmetic instructions so that all operands are always in registers.

5. The I-box forwards all operators to E-boxes with flags for synchronization.

6. Decode is 1 cycle.

7. Operand address formation takes 3 cycles plus a cycle to move the address to the memory interface. Address formation and decode will not be overlapped. However, the processor is capable of forming two addresses at the same time because it has two independent address adder mechanisms.

8. Memory references take 5 cycles including transfers to and from the processor/memory interface. Memory interface registers are unbuffered. There is one MFR and one MDR. However, transfers of storage addresses from MAR and values from MSR to memory may occur in parallel.

9. An ADD requires 3 cycles in its E-unit. A MULTIPLY requires 6 cycles in its E-unit. There are independent ADD and MULTIPLY E-boxes. Operand fetch is executed as an I-box function. Consequently LOAD instructions are not forwarded to E-boxes. Similarly STORE is executed by the I-box.

We operate on the following code, generated from A=(B*C) + (D+E).

| | |
|---|---|
| LOAD R1, B | B to register |
| LOAD R2, C | C to register |
| LOAD R3, D | D to register |
| LOAD R4, E | E to register |
| MPY R1, R2 | B*C |
| ADD R3, R4 | D+E |
| ADD R1, R3 | (B*C)+(D+E) |
| STORE R1, A | Result to A |

Let us look at Figure 17 to see how the list of instructions will execute in parallel. We start at the point where the address of the first instruction (LOAD R1) is in the IC.

Figure 17 is a Gantt Chart identical in form to Figures 9 and 10. Components are listed as the Y axis, time intervals as the X axis, and line segments for a component during an interval indicates that the component is busy. Line segments are keyed to an explanatory legend on the figure.

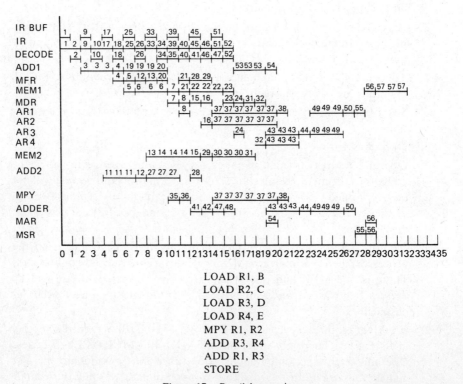

LOAD R1, B
LOAD R2, C
LOAD R3, D
LOAD R4, E
MPY R1, R2
ADD R3, R4
ADD R1, R3
STORE

**Figure 17.**   Parallel execution.

**Figure 17.**  (*Continued*)

1. LOAD R1, B to IR from IR buffer.
2. Decode LOAD R1, B.
3. Form address of B.
4. B address to MFR.
5. Request B from memory.
6. Memory fetches B.
7. B to MDR.
8. B from MDR to AR1.
9. LOAD R2, C to IR from IR buffer.
10. Decode LOAD R2, C.
11. Form address of C.
12. C address to MFR.
13. Request C from memory.
14. Memory fetches C.
15. C to MDR.
16. C from MDR to AR2.
17. LOAD R3, D to IR from IR buffer.
18. Decode LOAD R3, D.
19. Form address of D.
20. D address to MFR.
21. Request D from memory.
22. Memory fetches D.
23. D to MDR.
24. D from MDR to AR3.
25. LOAD R4, E to IR from IR buffer.
26. Decode LOAD R4, E.
27. From address of E.
28. E address to MFR.
29. Request E from memory.
30. Memory fetches E.
31. E to MDR.
32. E from MDR to AR4.
33. MPY to IR from IR buffer.
34. Decode MPY.
35. MPY to MPY E-unit.
36. E-unit continued decode.
37. MPY.
38. Product to AR1.
39. ADD R3, R4 to IR from IR buffer.
40. Decode ADD R3, R4.
41. ADD R3, R4 to ADD E-unit.
42. E-unit continued decode.
43. ADD R3, R4 execution.
44. Sum to R3.
45. ADD R1, R3 to IR from IR buffer.
46. Decode ADD R1, R3.
47. ADD R1, R3 to ADD E-unit.
48. E-unit continued decode.
49. ADD R1, R3.
50. Sum to R1.
51. STORE to IR from IR buffer.
52. Decode store.
53. Form A address.
54. A address to MAR.
55. Sum to MSR.
56. MAR, MSR (A and its address) to memory.
57. Memory places A.

The processor is able to complete the eight-instruction sequence in 32 elapsed cycles, an average of 4 cycles per instruction. This is not at all bad considering that a LOAD instruction alone requires 12 sequential cycles. So there is considerable overlap.

We notice that at various points we have multiple fetches, multiple fetches and the MULTIPLY, and the MULTIPLY and an ADD proceeding in parallel. Much of this overlap comes from the ability of the memory to effectively provide parallel response. Some delays are avoided because of the ability to form two addresses at the same time. In addition, the ability of the ADD E-unit to accept the second ADD before it completes the first frees the decode circuits by avoiding a delay for passing the instruction to the ADD E-unit.

There are some delays. In the interval $t_{10}$ to $t_{11}$ there is a memory reference delay because of our limit of two-way parallel response from memory. In the interval $t_{12}$ to $t_{14}$ the MPY E-unit must wait for C to arrive in its register. The ADD R1, R3 is similarly delayed, waiting for its operands. The STORE is delayed until the final ADD is complete.

One interesting note on the legend is the phrase "E-unit continued decode." This refers to a partitioning of function between the I-unit and the E-unit so that the I-unit does a minimum of decoding, just enough to determine who will execute the instruction, and further decoding is done in the E-unit. This permits the I-unit to get through with an instruction quickly and provides for potential parallel decode between E-units.

The behavior of this instruction list might be affected by memory requests for additional instructions. While all of this has been occurring, every time an instruction moved from the IR buffer to the IR another instruction was requested to fill the IR buffer position. This might result in some additional memory delays.

## 7.2.   Intersecting Memory References

We still have some problems with instruction sequencing. The previous list of instructions contained no common memory references. It is possible, however, that a program might contain the statements:

1.   C = A+B;
2.   B = G*D+E;

It is necessary that code generated for these statements assures that the value of B developed in (2) is placed in B after B is used to develop C in (1). We need some mechanism to assure the proper sequence of all instructions associated with (1) and (2).

The simplest way to accomplish this is to never allow a LOAD instruction to be executed when a STORE instruction is in execution and vice versa. To accomplish this a processor needs circuits that are capable of indicating that there is an outstanding STORE or LOAD. It need not remember the addresses of any outstanding instruction. The simplest design locks these instructions in the I-unit when a control delay is required.

This severe sequencing of fetch and store references would slow the machine in two ways. First, of course, it would delay the references to memory. Second, it would provide an instance where lookahead would be slowed and the parallel potential of the processor diminished. While the I-unit is waiting to release a STORE or LOAD instruction it cannot free the IR for further instruction preprocessing.

Since it is reasonable to expect that not all LOAD and STORE instructions will have identical addresses, it is desirable to apply synchronization rules only to those LOAD and STORE instructions that actually require sequencing to assure that a LOAD brings the correct value from a location.

A design to reduce delays makes use of the buffered memory interface registers to inspect whether a STORE or LOAD of the same address is currently in the memory interface buffer registers. If the address of a current STORE or LOAD is not found, then the instruction is forwarded to E-units. In this way references to memory may overlap except if a STORE finds its address in the MFR or a LOAD finds its address in the MAR. This coordination is necessary because of properties of memory access that cannot guarantee the rate at which MFR or MAR requests will be honored.

## 8. CODE EQUIVALENCE AND REARRANGEMENT

Logically unrelated instructions may be permuted in various ways on an instruction list. The ideal permutation is that which maps most closely onto the operational dynamics of the I-unit and E-units. We have suggested that an intelligent compiler might rearrange instructions in order to optimize the flow of independent instructions that may be executed in parallel.

The fundamental property of the relationships between independent instructions is that they are commutative. That is, they may be executed in any order in such a way that a machine status is achieved so that it is impossible to determine the actual order in which they have been executed. It is the observation of commutability that enables a compiler to rearrange code. The inference is drawn, as we have seen, from register references and memory references. To some extent it is possible for machine hardware itself to undertake limited dynamic rearrangements of the execution sequences of logically unrelated, commutative instructions. Thus an instruction list may not be executed in the order in which it is processed by the I-box. The ability of a processor to undertake local rearrangements of coding sequences has been called *chaotic execution*. The phrase does not mean, of course, that execution is chaotic, but that it is unpredictable within the limits of logical dependency. Chaotic execution allows a processor to make dynamic decisions about which instruction to execute next, depending on observation of machine status.

Chaotic execution may be designed for any E-box having a buffered internal DOR set. The I-box sends instructions appropriate for a particular E-box to that E-box rather than to a general DOR buffer area. This is appropriate for a heterogeneous E-unit machine where the I-unit can always determine which E-unit will execute an instruction. Operands are forwarded either to operand registers or to internal operand buffers in the E-unit so that at any time there is a population of operands and a population of operations to be performed.

The E-unit has sufficient circuits to determine when an instruction can be executed. It asks whether the operands for a particular instruction have arrived. If they have arrived it asks whether there is any reason not to execute this instruction. The reasons may be that it is dependent on the completion of a previous instruction or that the instruction is in a *zone of uncertainty* where it is not yet clear that this instruction will be executed. Some designs provide for the

I-unit to send instructions to an E-unit before the resolution of a branch. The E-unit must receive a signal that the proper branch has been taken before it will execute an instruction. The interval between receiving an instruction and receiving the delete or proceed signal for a branch instruction is called the zone of uncertainty.

When the E-unit inspects an instruction for execution and determines that it cannot be executed it may then visit a successor instruction. If conditions of independence, operand arrival, and certainty of execution (no zone of uncertainty) obtain, the second instruction may be executed before the first. The E-box always executes the first-found, ready-to-execute independent instruction. Thus a certain amount of local rearrangement of execution sequence is possible, although the sequential intent of the code is preserved.

We have not yet explained how it is possible for operands requested for a later instruction to arrive in the processor before operands requested for an earlier instruction. This will be made clear in Chapters 18 and 19. It can happen because of imperfection in the techniques that are used to avoid memory contention delays.

## 9. CHARACTERISTICS OF E-UNIT DESIGNS

This section undertakes to characterize various concepts relating to E-unit design. There are diverse approaches to the organization of E-units. They may be characterized as:

1. Single function, single level
2. Multifunction, single level
3. Single function, multilevel
4. Multifunction, multilevel

### 9.1.  Single-Level Units

A single-function, single-level execution unit is one that executes only one instruction within a single time interval that is not further divided. When an I-unit recognizes an instruction appropriate for a single-function E-unit, it sends a GO signal to the unit. The unit needs no other operator information since it only executes the one instruction. Such E-units require no DOR since they do no decoding of any sort. Operands are delivered to the units in operand registers within the unit. General-purpose registers could be used if some register reference convention existed so that addresses of registers used need not be sent to the E-unit.

A single-level, multifunction unit differs only in that it may execute a number of related instructions. Thus, instead of having units for each (potential) instruction there is a unit that does all ADDs or a unit that does all decimal arithmetic, or a unit that does all types of shifts. Such units must receive some directive code from decode and must make a determination themselves as to which function is required of them. It is for multifunction units of this type that we need a DOR in the execution unit.

Multifunction units must also wait for operands to be available in arithmetic registers, general operand buffers, or specialized buffers in their own unit.

The significant attribute of single-level units is that all of the instructions that they will execute will be executed sequentially. The unit will accept only one instruction at a time. The number of instructions that can be executed in any time period increases as the number of instructions that go to any particular unit is reduced and instructions are evenly spread around different execution units.

In the determination of how many execution units to build and how to group instructions within them, a designer must assess the degree to which the architecture of the machine will tend to provide instruction streams in which there is a high probability that instructions will be executable in parallel because there is no logical ordering between them. He must also assess the frequency of occurrence of instructions of different types.

Machines constructed of families of single-level, heterogeneous execution units are well-known in the industry. In such machines significant reductions in the time required to execute a stream of instructions that is properly formed and sequenced can be accomplished. Notice, however, that for machines of this class there may be lists of instructions that do not use the execution units well and do not achieve significant performance improvement as a result of the existence of multiple execution units.

### 9.2.  Multilevel Pipeline Units

The multilevel execution units introduce the notion of pipelining. The idea of pipelining is basically an extension of the idea we used to separate I and E functions. Looking at the execution of an instruction, we may determine that there are definable stages in the execution sequence and that an instruction may be moved through these stages in a way much like the way in which fluid flows through a pipe. Figure 18 shows an execution unit that breaks its instruction into stages.

An instruction flows through in fixed time sequences. While an instruction 1 is at stage 2, an instruction 2 can be at stage 1; when instruction 1 advances to stage 3, then instruction 2 advances to stage 2 and a third instruction can be admitted to the unit.

Let us say that the instruction being executed requires 6 machine cycles and that 1 cycle occurs at each stage. If the execution of six contiguous instructions were directed to the unit it would take 36 cycles if there were no pipelining. With pipelining, however, the first instruction would finish after 6 cycles, the second after 7 cycles, and the third after 8 cycles. The average number of cycles for an instruction executed by this unit would be 1.8 rather than 6.

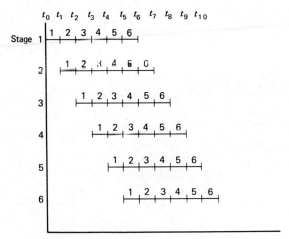

*Notes:*

1.  Numbers above line segments indicate instruction.
2.  First complete instruction at $t_6$, last at $t_{10}$, average instruction time: 11 cycles/6 inst. = 1.8.

**Figure 18.**   Pipeline.

The problem of a pipelined execution unit is getting work. A single-function, multilevel execution unit would be rather inefficiently used unless there were long sequences of the same instruction occurring in programs. It would not be impossible for a compiler to arrange for this to happen. For example, it might try to rearrange a program to group ADDs that were not related together. This would be hard work, however, and the constraints on doing a good job for various reasons of program flow would be severe.

The natural design for a pipelined unit is multifunction. There are some problems with a multifunction unit, however, in that the instructions it attempts to execute may have differing numbers of basic cycles. A MULTIPLY, for example, may have 10 cycles and an ADD 3 cycles. Also not all stages may be the same between grouped instructions. Consequently the designer of the pipelined unit faces the problem of alternative stages for different instructions. Also, where instruction stages are the same but differ in number, he may face a decision that requires slowing the simplest instruction to the rate at which it can process the most complex and passing it through null stages.

There is another consideration in pipeline design. When considering how many stages to define for an instruction, one may face a technology limit. It may be possible to break down an instruction into 20 stages, and, consequently, to define an enormous potential flow through the unit. However, the *latch time*, the time it takes to move an instruction from one stage to another, may be such that it is larger than the time it takes for each instruction stage.

The division into stages is limited by the relationship between stage processing time and latch time. A unit that spends too much time moving instructions from one stage to another relative to the time it is processing is not efficient. Designers of pipes are very interested in ways of improving the transition from stage to stage and in eliminating latch time as much as possible.

Machines with such an elaborate design are usually the glory of their time. Designs at this level are associated with rich architectures and a great interest in fast computation. We are even beyond the characteristics of the top of the line commercial machine of the IBM 3033 or IBM 3081 type and verging on the realm of the supercomputer. Historically, however, such designs have been used for machines marketed as general-purpose processors, and the problems they introduce to design are interesting for the student of machine organization.


## 10.  VECTOR MANIPULATION

An important design problem is achieving performance of a processor on very large, scientific programs where success is running a scientific program very quickly. The price/performance of a computer in this area may be less important than its raw performance across all problems. Or the processor may be price/performance efficient for the class of scientific work but not for commercial work that is brought to it only to occupy its time when critical work is not being done.

Those interested in this area tend to measure the performance of machines in *floating point operations per second* (*FLOPS*) and to concentrate on how the data structures associated with such calculations can be most effectively handled. These data structures are typically vectors and arrays of numbers.

An approach to the problem of high-speed calculation is to provide instructions in the architecture that manipulate vectors directly. The operands indicated by the address fields are not scalar variables but lists of numbers whose transformation in connection with other lists can be handled in parallel.

The vector arithmetic instructions are supported by a vector calculation E-unit that contains a number of input and output registers. When a vector operation is encountered in the instruction stream as many elements as can be handled by the vector execution unit are loaded into the unit from each vector. The arithmetic operator is then applied to all elements and the results are represented in the output registers.

Consider a simple example where we encounter the statement C=B*A where A, B, and C are declared vectors of length 8. Figure 19 shows a unit with eight element registers for A, B, and the output C. Elements of B and A are transferred as a block to the input registers. The MULTIPLY operation is applied to B1*A1 to produce a result in C1, to B2*A2 to produce a result in C2, and so on. The operation may be applied to elements sequentially, in parallel, or in a pipeline.

The efficiency of vector operations will be worthwhile even if absolutely parallel execution of the function is not undertaken. It is possible to approximate parallel execution with some form of pipelining and still achieve high

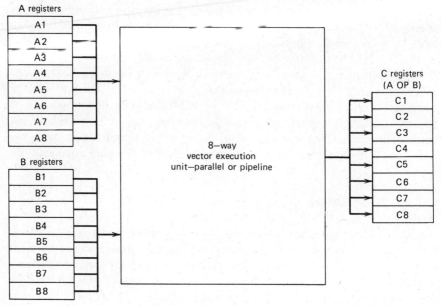

**Figure 19.**   Vector execution unit.

speeds. The problem of pipelining, getting enough work to fill the pipe, is solved because one instruction effectively provides a long list of operands. The burden of multiple instruction fetches is removed as is the burden of index register modification and counting necessary when vector operations are coded as loops.

In programs that basically calculate on vector and array structures where each arithmetic instruction represents a significant portion of the work of the machine, many loads and stores are eliminated and the density of arithmetic instructions justifies the design. The population of programs that are basically loops within loops is sufficiently large so that the approach is more than a special-purpose approach. Some commercial applications can be conceived of and programmed to assume the characteristics of vector-oriented manipulations.

An alternate approach to fast vector operations using parallel execution is to use homogeneous, highly intelligent E-boxes so that every E-box executes any instruction in the system. Each E-box has a private memory space which contains a subset of the elements of the vectors that are to be manipulated. An instruction broadcast unit (I-box) broadcasts an instruction with a list of addresses to all execution units which then operate that instruction on local data. Machines of this type are called *single instruction multiple data* (*SIMD*) machines. Members of this type are the SOLOMON (possibly the first) and the ILIAC IV.

A problem with this type of machine organization is organizing work so that each of the execution elements contributes significantly to the progress of the work. This involves rather complex analysis of memory layout and problem definition. In addition, methods for using machines of this structure to compile for themselves have never matured and it is commonly necessary to compile for them on other machines. Typically the vector processors of this type are attached to more general-purpose machines which communicate with the instruction broadcaster of the vector processor across a shared drum. When the general-purpose processor encounters a portion of the program suitable for vector processing it sends it to the instruction broadcast unit for distribution to execution elements.

## 11. MULTIPLE I-STREAM MACHINES

The goal of some designers is to create a machine that will execute effectively in environments where there may be no very large program dominating a machine. It is common in commercial environments to *multiprogram* or *time-share* systems across a population of users or programs. Each program is allocated some part of the resources of the machine at different intervals by the action of the operating system. This joint use may be supported by only one multiplexed uniprocessor or by a family of identical processors with common access to memory. Such machines are called *multiprocessors* and are part of a class of machines called *multiple instruction multiple data* (*MIMD*) machines.

There are various forms of multiprocessor designs, with different degrees of resource sharing, that undertake to support multiple programs across a set of processors within a single system.

In Figure 20, the processors of A share memory but have private I/O populations. In B, processors share both memory and a common I/O pool. In C, processors do not share memory but are bus interconnected and talk to each other on a message basis, processor to processor. C is not a classical multiprocessor but a form of *distributed processor* that is emerging from advances in technology. What these pictures have in common is that the processing units are complete packages of I and E function, perfectly complete processing units capable of stand alone operation if attention has been given to reconfiguration and reliability features.

Some designers have been interested in the notion of *multiple I-stream* machines where a population of I-boxes shares a population of heterogeneous execution units. Such a system is represented by Figure 21.

The figure shows four I-units, each with a private instruction counter and instruction register. These units interface to memory for instruction fetch. Processed instructions are passed to a common system DOR buffer. Any E-box takes instructions of appropriate type from this buffer. It is necessary that instructions placed in the DOR contain an identifying code indicating the I-unit from which the instruction has come. This identification is used in various ways. In particular it is used to index into a Regselect (Register Selection) array that indicates the partitioning of the pool of operand registers among I-units. A

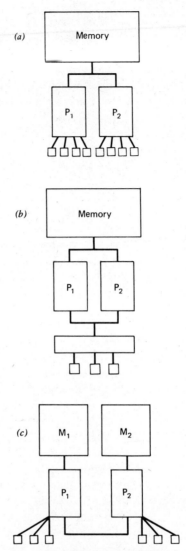

**Figure 20.** Multiprocessor organizations: (*a*) memory shared, private I/O; (*b*) memory shared, pooled I/O; (*c*) no memory share, private I/O, interprocessor bus.

feature of the design of Figure 21 is dynamic partitioning of the operand registers between the I-streams active in the machine. Register references sent to an E-box are relative to the lowest numbered register assigned to an I-unit.

There have been many variations in proposed and real machines of this type. Variations include how much circuitry is unique to each I-unit and how much is shared. Decode circuits, for example, may be shared among I-units or duplicated in each one. Private sets of operand registers organized as multiple register sets may be associated with each I-unit so that dynamic partitioning of the register set is neither necessary nor possible.

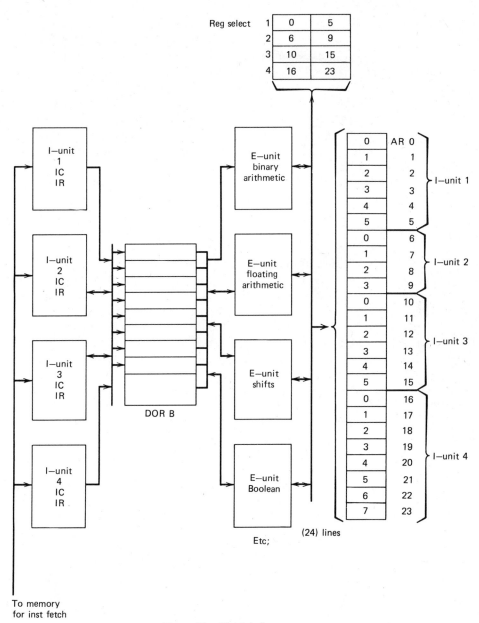

**Figure 21.** Multiple I-stream.

The notion of a multiple I-stream machine of the Figure 21 type comes from consideration of the difficulty of finding instruction sequences that will use the execution units in a balanced fashion in a single program. The expenses of instruction rearrangement and the complexity of designs that try to keep the multiple execution units busy within the context of a single program seem unattractive to those interested in this class of machine.

The contention is made that the probability of finding logically unrelated instructions is higher if there is more than one active instruction stream. Therefore a network of I-boxes connected to a set of E-boxes will have greater throughput than drops of efficiency "squeezed" out of a single program. Instructions from different instruction streams are largely unrelated and can flow into available E-units to provide a greater balance of heterogeneous instructions executable in parallel than can come from one program.

The contention is also made that multiprogramming techniques that rapidly shift from one program to another will diminish the effectiveness of pipelines. The pipeline may have to be drained whenever a program switch is made and a pipeline processor will not perform most efficiently in a multiprogramming environment. This is true, for example, in a machine that uses condition code branch logic. Instructions in the pipeline may affect the condition code and no instructions can enter the pipe from a new program until the last instruction from the old program has cleared the pipe. Consequently it is better to organize for continuous multiple program execution around multiple, non-pipelined units in the opinion of some designers.

A multiple I-stream machine might contain a list of instructions in a set of multiple ICs and IRs where each IC and IR is dedicated to a specific stream. The processor submits instructions to the execution units in a round robin manner so that at any time the population of instructions under execution is spread across the population of active I-streams. In this way the execution units are kept busy with sets of unrelated instructions.

One of the difficulties in this approach has been the reduction in cost of circuits which makes the economics of sharing E-boxes less important than it was. However, there is some reason for continued interest in this notion cast in a slightly different way.

One effect of the multiple I-stream capability is to provide a program switching capability for multiprogram environments at close to 0 hardware time cost. Thus a machine with multiple ICs and IRs in connection with private register sets could program switch very quickly and support multiple user environments very well. Such a design would be effective even in smaller machines that do not have specialized multiple execution units. The goal is to minimize the time spent in the operating system and not to maximize execution unit utilization. Program switching might well be on an instruction-to-instruction basis or at some slower rate that allows bursts of instructions to be performed by one program before the IC of another is given control of circuits.

## QUESTIONS

1. What are two requirements for executing instructions in parallel?

2. What are two ways of grouping functions for E-units?

3. How is operand forwarding achieved in an architecture that allows memory references in functional instructions?

4. What is a basic instruction sequencing scheme?

5. How may the basic scheme be improved?

6. Draw a Gantt Chart showing how a lookahead machine would perform the instruction list of Figure 17.

7. What is chaotic execution?

8. What is pipelining?

9. What is the basic nature of vector operations in a vector E-unit?

10. What is the rationale for a multiple I-stream machine?

**Chapter 18**

# Memory Organization

## 1. BASIC GOALS OF MEMORY DESIGN

This chapter discusses memory design techniques and the interconnection of memories and processors.

There are a number of fundamental issues that associate with the design of memory. In general the design of a memory must consider:

1. How much memory is required for processors of different speeds.
2. Minimizing contention between memory references.
3. Minimizing the impact of the processor speed/memory speed imbalance.

The amount of memory affects the extent to which the speed of the system is bound by the speed of the memory rather than by the speed of I/O. Reducing contention is an attempt to avoid having a processor wait for references. Minimizing the impact of speed imbalance is an attempt to find methods by which the system can run at processor rather than at memory speeds.

## 2. ENOUGH MEMORY

Despite the complexity of determining how much memory a system should have, there is a simple underlying concept. A fast processor should have more memory than a slow processor. A fast processor can execute instructions and transform data at a rate that may require constant I/O operations to bring in more instructions or more data. A slow processor will tolerate smaller memories because it uses instructions and data at a slower rate. This extends the period of time during which the contents of memory are current. This reduces the rate at which I/O must be done. I/O is much slower than memory, which is in turn slower than processor logic. High levels of I/O activity introduce the possibility that there will be intervals of time during which the processor has nothing to do because it is waiting for new instructions or data.

238

It is difficult to provide a single value for a general processor speed/memory size ratio. Parameters that affect the amount of memory that a system requires include:

1. The speed of I/O devices.
2. The speed of pathways that connect to I/O devices.
3. The spread of data and instructions across devices.
4. The number of I/O devices that can be read/written in parallel.
5. The expected number of programs in a multiprogramming mix.
6. The expected application program size.
7. The expected number of instructions to process a transaction or a record in the processor.
8. The expected number of instruction executions between I/O operations.
9. The size required for underlying systems software. This includes resident portions of the operating system, and subsystems, and so on.
10. The efficiency of the architecture in representing data and instructions in minimum address space.

The speed of the I/O system and its potential parallelism determine the rate at which data may flow into the memory. If access rates are high, if contention for devices is low, and if many devices may operate in parallel then less memory may be required because of the ability to refresh the contents of memory quickly. However, the memory must be big enough to receive large amounts of data from different devices at the same time. The organization of the memory must also provide for parallel access from multiple devices.

The efficiency of the architecture determines the amount of memory space needed to hold instructions and data. As we have seen in Chapter 12, one figure of merit for an architecture is how well its instructions and data can be compressed.

The size of underlying software determines how much memory must be given over to vendor provided software packages that provide services to application programs.

The problem of determining how much memory to put on a processor is common to both designers and users of computing equipment. Designers must determine the minimum, maximum, and intermediate memory sizes to make available. Users must choose what memory size is required for their configuration.

## 3.  MEMORY ORGANIZATION

The chapters on lookahead and parallel execution assumed a memory that could satisfy more than one reference at a time. The need for this feature was to reduce delays due to contention for memory use. A memory that can satisfy multiple requests will reduce the time of a reference to the memory response time. Time spent waiting for access to memory is minimized.

The design technique for achieving multiple memory response is to organize the address space of an architecture into distinct units of memory, each of which can respond in parallel.

The actual physical locations of a memory may be organized in a number of different ways. Two basic approaches are *interleaving* and *banking*. Both these techniques divide the address space of physical memory into a family of memory units with a distribution of addresses across each unit.

### 3.1.  Banking

The basic memory design technique for reducing contention is called banking. In banking the memory is divided into banks in such a way that the upper bits of an address in the architecture are used to select a memory bank that contains addresses within a certain continuous range.

If the upper 2 bits of an 8-bit address are used to select a bank of memory, then:

| Full Address | Memory Bank | Relative Position in Bank |
|---|---|---|
| 0010110 (22) | 00 | 22 |
| 0110110 (54) | 01 | 22 |
| 1010110 (86) | 10 | 22 |
| 1110110 (118) | 11 | 22 |

This small example shows a memory whose entire range of addressability is 128 locations. These locations are divided into four memory banks, each with 32 locations. The upper 2 bits of the address select memory banks 00, 01, 10, and 11. The lower 6 bits indicate a location within each memory bank. The largest address within each bank is 31. The spread of memory locations between banks is therefore:

| Bank | Addresses |
|---|---|
| 00 | 00–31 |
| 01 | 32–63 |
| 10 | 64–95 |
| 11 | 96–127 |

Naturally we are usually interested in much larger address spaces, but the organizing principle still applies. The number of banks is determined by the number of high-order bits of address used for bank selection. The size of each bank is determined by the number of remaining bits used to determine displacement in the bank.

The effect of banking is to provide concurrent memory service for addresses that occur in different banks. Thus if instructions are in location 0 to 128K and data are in 128K to 256K, then instruction fetches and data fetches or stores may proceed in parallel. This memory organization is what leads to the parallel properties of the memory we assumed in the lookahead and parallel execution chapters.

Notice that the probability of a contention (reference to the same bank) decreases as the number of banks increases and the addresses held in each bank decreases.

## 3.2. Interleaving to Speed Memory Response

In interleaving the lower bits of the address are used for memory bank selection. If just 1 bit was used there would be two memory banks, one holding all even addresses and one holding all odd addresses. If 2 bits were used, there would be four memory banks as follows:

| Bank | Addresses | Interval |
|------|-----------|----------|
| 00 | 0000, 0004, 0008, etc. | 0004 |
| 01 | 0001, 0005, 0009, etc. | 0004 |
| 10 | 0002, 0006, 0010, etc. | 0004 |
| 11 | 0003, 0007, 0011, etc. | 0004 |

The number of low-order bits determines the number of banks and the spread of interleaved addresses within a bank.

The primary design goal of interleaving is to address the processor/memory speed imbalance. It involves a number of memory banks operating in parallel in the service of a single request. For example, a LOAD MULTIPLE instruction that desires to load four operand registers from four consecutive locations in memory could take a location from each bank and complete the operation in nearly the same time it would take to load a single register.

Interleaving may have a tendency to increase the contention in a system because it maximizes the number of memory banks that are involved in a single reference to memory. There is a trade-off to be made in designing processor/memory relationships between minimizing contention and speeding a single operation. A system that has trivial lookahead or parallel potential will profit most from the interleave design because consecutive memory references are in any case executed sequentially. The processor is not designed to fetch instructions and operands in parallel and consequently will not profit from banking. However, the interleave may reduce the amount of time required for any fetch or store. The analysis of the proper trade-off between banking and interleaving in systems with parallel capability or extended lookahead is very complex as it is dependent on various concepts of work load characteristics.

Many systems combine the two techniques, offering, for example, odd/even interleave within banks that are addressed through the high-order bits.

A byte addressable machine introduces another factor in memory bank organization. As we have seen in earlier chapters, a byte machine may have addressable units of multiple bytes. Halfwords, words, doublewords, and so on that exist as addressable units in many instructions. In the initial example of interleaving we suggested that each byte would be in a different memory bank. It is not necessary that this be true and a different *granularity* of address distribution may be designed. It is possible, for example, to place halfwords in each bank. Then each bank will contain addresses on halfword boundaries. It is even possible to choose a granularity that is not in the architecture. In the following example, the unit of distribution is a 4-byte word and there are four banks.

| Bank | Addresses |
|------|-----------|
| 00 | (0000–0003), (0016–0019), etc. |
| 01 | (0004–0007), (0020–0023), etc. |
| 10 | (0008–0011), (0024–0027), etc. |
| 11 | (0012–0015), (0028–0031), etc. |

### 4.  MEMORY LOOKAHEAD

In many larger systems the number of addressable units transferred between processor and memory is not reflected in the architecture. Some number of words or bytes that is not known to the architecture is manipulated on a memory reference. Thus a unit of 8-byte doublewords (64 bytes) may be transferred between memory and processor even though there is no instruction in the processor that can reference a 64-byte structure directly and there is no operand register in the processor that can hold a 64-byte unit.

The reason for transferring units of data that exceed architectural specification is an anticipation that data near an addressed location will be needed in the near future. If the memory is interleaved, the transfer of additional data may not take significantly more time and may eliminate additional memory references later on. There must of course be some staging or buffer area for the additional data that is moved to the processor.

In various models of the 370 generic architecture a reference to memory will result in the parallel transfer of very large collections of bytes that include the address requested and a set of neighboring addresses. The unit of memory is a doubleword in all high-performance models of the architecture, the interleave varies from four-way to eight-way, with doublewords interleaved through modules of memory that contribute to a transfer of data beyond that defined in the architecture.

The transfer of additional data may reduce memory contention on an interleaved memory. An example of this can be given with the notion of IR buffering that we have already discussed. In filling the IR buffer area, it is possible to reduce the number of times that a processor must go to memory by bringing in more than the one instruction requested. Thus if the system had four memory units, 32 bit instructions, and a 32-bit width in each memory, four instructions could be brought to the IR buffers on one memory reference. Contention might be reduced because memory is referenced only once per four instructions.

The effect of a bulk data transfer is to maximize the data rate between memory and processor and hope that a maximum amount of the data transferred will be used. When these data are instructions there is, in fact, a high probability that this will be so. The probability that operands transferred in bulk will all be used is less certain.

## 5.   MEMORY PARTITIONING TECHNIQUES AND PROCESSOR LOOKAHEAD

In Chapter 16 we assumed that there would be no memory contention delays to interfere with processor lookahead. Such a goal is clearly supported by a banking design. The effect that interleave will have on memory contention is less clear as it attempts to make each interaction with memory more productive. A few comments on the relationship between lookahead and memory organizations are appropriate here. Basically, interleave seems to increase contention because it involves all memory units in each reference.

The goal of design is to minimize the time a processor waits for the completion of a memory function. By interleaving and by using super-architectural transfers that bring in more data than were actually requested, the effective speed of the memory may be increased. When multiple instructions are fetched from memory the average speed of delivering each instruction is increased and contention is reduced. Contention is reduced because the number of times the processor goes to memory for instructions is reduced. The processor has a pool of instructions available outside of the memory. Thus it has faster access to each instruction and makes fewer memory references for instructions. Processor lookahead is served by the processor's ability to search forward on the already gathered pool of instructions.

Data references, which may also contain more information than requested in the hope that many usable data units are transferred, are also helped by interleaved super-architectural transfer. Even if multiple data elements are not used, data references have a reduced contention problem because the level of instruction references are reduced. Without super-architectural memory lookahead, interleave increases contention. With it, it seems possible that it is decreased relative to banking designs.

## 6.   LIMITS ON MEMORY PARTITIONING

The use of memory interleave faces some technology issues that derive from the success that the industry is achieving in placing more and more memory locations in smaller and smaller physical packages. When a basic memory element might contain 4096 words and units of this size were relatively expensive the cost of separately packaging and connecting many memory banks was modest compared to the cost of the memory. Since memory capacity per unit was small, the addressing granularity of the memory units could be small. Thus it would have been profitable to have many smaller memory banks and maximize interleave spread or minimize banking granularity.

64K bytes of information can now be placed on a chip. A unit of memory, consisting of a number of chips, may contain millions of bytes of data. The cost per unit of stored memory has gotten to be very low for memories that operate in the range of hundreds of nanoseconds. Thus it is more efficient to package large amounts of memory as a unit. The granularity of the address spread becomes much larger. Only large processors using many megabytes of memory can use separate memory banks economically.

This memory packaging phenomenon implies something about the use of lookahead for achieving speeds in processors with memories in the 1 million-byte range and below. Significant processor lookahead must be supported by making very large transfers on one request from the same memory unit. Thus wider data paths may be effective in smaller machines. The single memory delivers more usable instruction groups from a single reference to memory. More usable data may also be delivered. Effective lookahead on smaller machines, therefore, seems to be more dependent on larger local buffers within the processors than on interleave.

## 7. MEMORY TIMES

The time required by a memory to perform operations may depend on the function to be performed and the pattern of memory references. There are a number of particular times associated with memory behavior. All memories have *access times* and *cycle times* for read and write operations. Whether the access and cycle times for read and write are identical depends on the basic technology used to construct the memory.

Access time is the time required to read or write data at a memory location. A memory may have, for example, a 300-nanosecond access time. This means that, from the time that a request for function is accepted until the operation is complete, 300 nanoseconds have passed. It is possible, however, that another period of time may elapse before the memory is available for another operation. Thus cycle time consists of the access time plus any additional time after the completion of a function until the memory is available for another function. For example, some memories require a restore period after a read operation. The act of reading out the data has changed the status of memory, and circuits must be restored before the memory is again usable. Some memories take different amounts of time to write and read data. The actual time for a memory operation then depends on whether a bank addressed for a previous operation is addressed immediately for a successor operation and on the pattern of reads and writes addressed to the system.

One effect of memory partitioning in interleaved or banking form is to reduce delays in consecutive references to memory due to restoration time. Banking does this by undertaking to spread consecutive memory references (for an instruction followed by an operand fetch) to different banks so that the first referenced bank may restore itself without causing delay. Interleaving does this by attempting to provide a set of addresses on each reference so that additional references may be delayed until restore cycles are complete.

## 8.   A MEMORY/PROCESSOR INTERCONNECT ORGANIZATION

An issue distinct from the organization of main memory is the means by which a memory and a processor are interconnected. There are various approaches to this interconnection. Figure 22 shows a design that involves a processor, a set of memory banks, and a unit called a *primary storage control unit* (*PSCU*). The PSCU is the interface between the processor and memory. It is the vehicle through which all registers that hold requests for data or data to be transferred communicate with memory.

The reader will notice that the memory interface registers shown in the processor differ somewhat from the interface design we discussed earlier. In place of a register that holds memory requests (MFR), a register that holds data returned from the memory (MDR), a register that holds the address of data to be stored (MAR), and a register to hold data to be stored (MSR), the pictured processor shows an *operator request register* (*OPR*), an *operator return register* (*OPT*), an *operand request register* (*NDR*), an *operand return register* (*NDT*), and an *operand storage request register*. Notice that the operand storage request register differs from the other registers in that it has two elements, an address (SAR) and a value (SVR). Any request to store in memory must provide the value to be stored as well as the address into which it is to be stored.

In this particular processor there is a single I-box and a single E-box. We will ascribe to the I-box the responsibility for making all memory requests. The I-box, then, is a user of the operator request register, the operand request register, the operator return register, and the operand storage request register. The E-box is responsible for transferring data from the operand return register to various arithmetic registers in the machine. The I-box will place operand values as well as address values in the operand storage request register.

The PSCU is responsible for coordinating memory functions and for providing a path from the processor to the memory. The PSCU is sensitive to changes in the contents of any request register, on one side, and to the presentation of data from the memory on the other. Whenever a value is loaded into a request register a signal is sent to the PSCU. The PSCU responds with a request for the contents of the appropriate register. These data are sent on data lines that connect the PSCU with the processor. When the PSCU receives an address it determines the spread of addresses across memory units that will be involved. It determines whether these units are currently free to be accessed. If so, PSCU

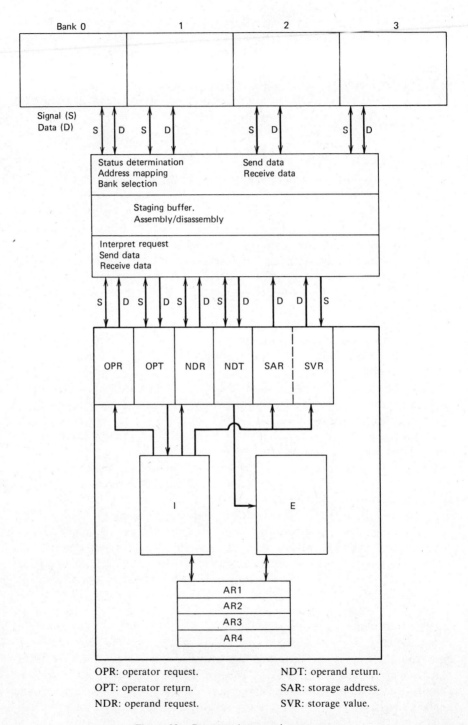

Figure 22.    Processor/memory interconnect.

OPR: operator request.    NDT: operand return.

OPT: operator return.     SAR: storage address.

NDR: operand request.     SVR: storage value.

passes the necessary addresses, and the necessary data when appropriate, to the units.

The memories may respond asynchronously in arbitrary order. The PSCU must form the completed unit of transfer in registers of its own and then transfer the collection of data, as a group, into the appropriate return register.

The lines between the PSCU and each memory bank indicate that a transfer of data from each memory bank to the PSCU may occur in parallel over those lines. Each line represents a data flow and a control flow, both of some width.

The natural width of a data line is the size of the unit that the memory is asked to transfer or receive. A byte organized memory unit would typically have an 8-bit wide data line. A 32-bit word organized memory might have a 32-bit wide data path. However, it is common in small or intermediate size machines for the data path to be narrower than the architected structure. Thus a request for the fetch of a 32-bit word contained in a single memory unit may result in two transfers of 16 bits each.

The width of the control line varies. A control line may be only 1 bit wide and used only to attract the attention of the units at either side. With a single-bit control line, control information is sent as data over a data line. The units know it is control data either because of some identifying tag or because of a sequencing mechanism that knows that at a particular stage of operation all data is to be interpreted as control information. Control lines, however, may be as wide as the data structures that are required to represent control information. Thus all function control information can be sent with a single transmission over the control line.

The actual protocols of the PSCU, memory interaction, depend on basic decisions about the width of the control line, the width of the data line, and the use of control lines or data lines to transfer control information. An interface might use a control line of 1 bit that is used by the PSCU to determine if the memory is free. The memory sends busy or free signals to the PSCU whenever its status changes. When the PSCU gets a memory free signal it determines whether it has a function for that memory to perform. If it wants the memory to store some data, for example, it undertakes a series of interactions with the memory to bring about the write.

Initially the PSCU sends a 1-bit signal on the control line indicating that a memory function request is about to take place. When the memory responds with a signal, the PSCU begins to transmit packets across the data line until all address and function data have been passed. The memory acknowledges receipt of each packet with a signal. If the function is a write to memory, the PSCU sends packets of data to be stored in a similar fashion, the memory acknowledging the receipt of each packet. When the storage is complete, the memory sends the PSCU a complete signal.

In Figure 22, instructions and operands have separate lines to the processor request and return registers from PSCU. In addition there is a separate storage line to the PSCU from SAR and SVR. This implies an ability to flow operands and instructions in parallel. The actual configuration of lines between processor and PSCU will depend on the details of memory organization as well as processor organization. An interleaved or banking memory system might be configured differently at the processor to PSCU interface.

The fundamental feature of the design of a PSCU is that it is capable of sustaining an interaction with the memory on one side and the processor on the other in parallel.

## 9. ALTERNATIVE INTERCONNECTION ORGANIZATION

Figure 23 shows another way of interconnecting processors and memories. It is called *multiport interconnection* and does not involve a PSCU.

The functions of the PSCU are split between the processor and the memory which has an intelligent component that can sequence requests, resolve contention, and so on. In Figure 23, each processor has four ports that go to memory and each memory has four ports that can go to processors. Each processor has

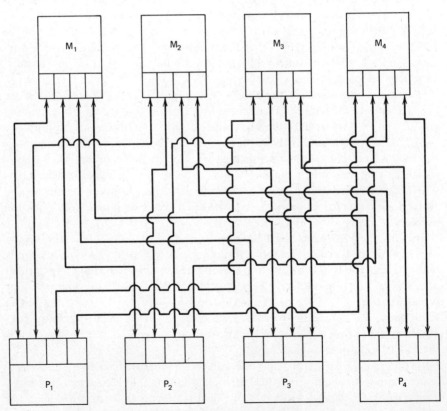

**Figure 23.** Multiport interconnect.

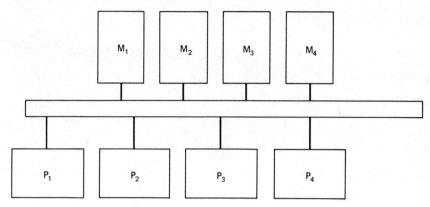

**Figure 24.**   Single bus interconnect.

a path to each memory so that contention can only occur at the memory. If the memory is free there will be no delays along any pathways. The number of interconnecting circuits is P (number of processors) × M (number of memory banks).

Readers who are familiar with some networking concepts will recognize the PSCU configuration as a *star* and the multiport configuration as a *mesh*. Since memory interconnections are networks, there are numerous variations in their design. Figure 24 shows each line coming from a port of a processor joining a single bus that connects to memories. This suggests possible contention on the main memory interconnect bus. The impact on performance depends on the relative speed of the bus compared to the rate at which processors can make memory references. The actual mechanism for resolving contention involves arbitration circuits that request the bus when it is free, the first to request a free bus getting control of it for some period of time. This time will be, at minimum, the time it takes to complete one memory operation. Bulk moves to or from memory may be interrupted, and the bus given to some other unit and completed over an extended period of time. This implies some elaboration of staging and sequencing in the processors and the memory.

Some priority scheme for memory bus use may be used instead of first come/first served. Particular processors may be given particular priorities, frequently as a function of where they are in the physical structure of the bus. Yet another way of handling multiple use of the bus is to define fixed periods of time during which each processor can use the bus. This *fixed time division multiplexing* avoids contention by sequencing in a round robin fashion among processors.

Notice that in both multiport pictures all processors may connect with all memories and that, as a result, the entire address space is equally available to all processors. It would be possible to constrain the addressability of any processor by not connecting it to the spaces available from a particular memory. This would only be meaningful in high-bit interleaved banking systems.

QUESTIONS

1. What are the basic goals of memory design?

2. What factors influence how much memory a processor should have?

3. What is high-order bit banking?

4. What is interleave?

5. What is the possible effect of interleave on memory contention?

6. What are the functions of a PSCU?

7. What is a multiport interconnect?

# Chapter 19

# Memory
# Hierarchies

## 1. NOTION OF HIERARCHY

Throughout the book one has been alluding to the possibility that there may be storage space in a processor that can be used as a buffer between the processor and the memory. We made an early passing reference to caches and showed instruction and operand buffering with the IR, DOR, and E-unit operand buffers.

The notion of memory hierarchy exists in a rudimentary form with all processors that have significant sets of registers. The use of the registers is to provide storage and buffering spaces so that data can be accessed and manipulated without reference to main memory.

At various times one referred to the processor/memory speed imbalance that occurs with fast processors. The imbalance means that a processor that can accomplish an instruction in a given number of machine cycles is delayed because of the large number of machine cycles that elapse on a reference to memory. Interleaving attempts to address this problem, as does memory look-ahead with associated buffering. Further design to improve system speed involves some extensions and elaboration on the notion of buffering.

The idea of memory hierarchy is to provide memories of various speeds and sizes at different points in the structure of a system. These memories are increasingly large and slow as one moves away from the processor. The mode of addressing may change from memory level to memory level. Figure 25 shows this concept.

The figure shows a processor with a set of internal IR and DOR buffers and a set of arithmetic registers. The processor also has a small area in which to hold units of transfer that exceed architectural units. Beyond the processor is a primary memory with a 300-nanosecond cycle time configured at 4 million (mega) bytes. Behind this memory is a bulk static memory with a 1-microsecond cycle and 20 megabytes. Behind this secondary memory is a drum, behind the drum are disks, and behind the disks some kind of archival device.

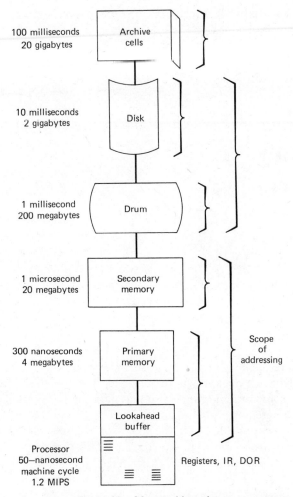

**Figure 25.** Memory hierarchy.

The brackets in the figure indicate points where the addressing conventions might change. In various approaches to hierarchy, automatic address translation occurs at different points. Where two levels of hierarchy are connected by automatic address translation, program addresses occur in terms of the closer unit and are automatically mapped into the further unit so that addressing details are hidden to the program. When such translation mechanisms do not exist between levels of the hierarchy, the addressing conventions of the farther unit are visible to a program. This is typically true of spaces in disk units that are commonly referred to using I/O commands providing unit, cylinder, surface, and track addresses.

Each of the rotating storage devices may have some static storage of its own. For example, a disk system may have some local memory that it uses as a staging and lookahead area. In addition there may be small processors associated with the storage devices that control the behavior of the unit.

This chapter discusses only those portions of hierarchy that are not referenced by I/O instructions. The line between storage device and memory is becoming fuzzy as hardware and software technology seem to be moving toward a concept of addressability that spans the total storage capacity of a system with automatic address translation between all levels and paths from level to level that do not move through the processor primary memory.

## 2.  INSTRUCTION BUFFERS

We now look more closely at instruction buffering, first discussed as back-up IRs. A buffer that holds instructions fetched from memory as a result of I-box lookahead is commonly called an *instruction buffer*. We investigate various means of filling and controlling the contents of such buffers.

The intent of the element is to provide a repository of instructions that are available to the processor when it wishes to undertake instruction preparation. The size of this memory may be in the area of 4 to 64 instructions. The unit is matched for speed with the logic of the processor.

The instruction buffer is commonly an implicit buffer in that there are no instructions that specifically call for the loading of instructions in the buffer. Instructions move to the instruction buffer as a result of I-box sequencing and IC instruction address generation. Instructions may be moved one at a time or in groups. In a lookahead or parallel machine the addresses of instructions some distance beyond the current instruction being executed can be computed in parallel with other functions and a continuous stream of instructions may be kept flowing into an instruction buffer.

Figure 26 shows a machine with a four-instruction buffer. The table below shows the state of the system at the time the STORE instruction of Figure 26 enters execution. We are making a simplifying assumption that only one instruction may be in I-phase or E-phase at a time. This need not be true.

| Instruction | Memory Location | Buffer Location | Status |
|---|---|---|---|
| STORE R1, X | 1003 | 3 | In execution. |
| LOAD R1, A | 1004 | 0 | In I-time. |
| LOAD R2, B | 1005 | 1 | Waiting in buffer. |
| ADD R1, R2 | 1006 | 2 | Waiting in buffer. |
| MPY R1, C | 1007 | X | On way to buffer. |
| STORE R1, D | 1008 | X | About to be requested by IC. |
| LOAD R1, E | 1009 | X | Not yet requested. |
| LOAD R2, F | 1010 | X | Not yet requested. |
| DIV R1, R2 | 1011 | X | Not yet requested. |

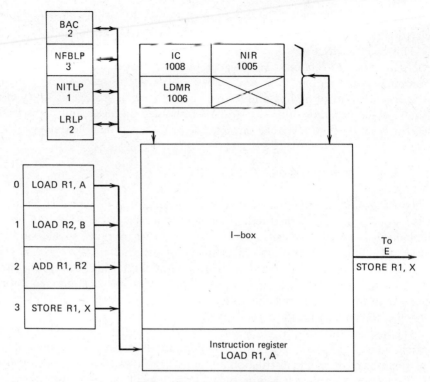

**Figure 26.**   Instruction buffering.

This status of the system is represented in hardware by the buffer and a set of control registers. A reasonable set of control registers for this simple example is as follows:

1. *Next Free Buffer Location Pointer (NFBLP).*   A 2-bit register that points to the next location in the buffer that should receive an instruction.
2. *Next I-Time Location Pointer (NITLP).*   A 2-bit register that points to the buffer location that will next provide an instruction for I-time.
3. *Last Receiving Location Pointer (LRLP).*   A 2-bit register that points to the last location to receive an instruction.
4. *Buffer Available Counter (BAC).*   A 2-bit register that holds a value representing the number of free buffer spaces.
5. *IC.*   The instruction counter that provides the address of the next instruction to be fetched from memory.
6. *Last Delivered Memory Register (LDMR).*   An address register holding the address of the instruction that was last delivered to the buffer.
7. *Next in I-time Register (NIR).*   An address register holding the address of the next instruction that will be delivered to IR.

The status of these registers, representing the present condition of the machine, is as follows:

| Register | Value | Meaning |
|---|---|---|
| NFBLP | 3 | Next instruction will overlay STORE in buffer entry 3. |
| NITLP | 1 | Next instruction to enter I-time (LOAD R2, B) is in buffer entry 1. |
| LRLP | 2 | Last instruction delivered to buffer (ADD R1, R2) is in entry 2. |
| BAC | 2 | There are two available locations in the buffer. They are 3 and 0. LOAD R1, A has just been moved from 0. |
| IC | 1008 | Address of STORE about to be requested. |
| LDMR | 1006 | Address of ADD last delivered to buffer. |
| NIR | 1005 | Address of LOAD R2, B next to go to I-time. |

The dynamics of the mechanism is as follows:

### When an I-Time Completes:

1. Compare NIR with LDMR. If NIR is smaller then the next instruction is in the buffer. If the instruction is not in the buffer wait for the instruction to arrive. If the instruction is in the buffer, transfer it to IR and
2. Increase BAC to show that a buffer space has become free.
3. Increase NFBLP to point to free buffer entry just vacated.
4. Advance NITLP to point to buffer entry containing next instruction.
5. Advance NIR to point to the address of the next instruction to enter I-time.

These functions must be done as much as possible in parallel hardware to minimize elapsed time.

### When an Instruction Arrives at the Buffer:

1. Increment LDMR to show last delivered instruction address.
2. Increase and, if necessary, adjust LRLP for wraparound to show last used buffer entry.
3. Decrement BAC. If BAC is 0 suspend instruction fetch. If BAC is not 0 then increase IC and fetch next instruction. When an instruction is withdrawn from the buffer and BAC is increased a start fetch signal may be generated. Proper synchronization of BAC increment and decrement must be assured.

The above functions should be implemented in hardware with sufficient parallelism so that a significant number of machine cycles do not add to instruction timing.

The buffered processor will outperform an unbuffered processor to the extent that it finds instructions in the buffer. This is determined by the relative rates of instruction fetch and instruction withdrawal that determine the probability that BAC will be 0.

Instructions may be brought into the buffer in groups. With an interleaved memory an instruction reference can cause the transfer of the referenced instruction and some set of successor instructions in a time period roughly the same as that needed for the delivery of a single instruction. Thus one reference to memory will bring forth a group of instructions, reducing the time to get instructions and reducing the number of individual memory references required to get part of an instruction stream.

Group instruction transfer will change the details of the manipulation of the various registers that control and indicate buffer contents. In addition, if a single word contains multiple instructions an additional staging area before the IR may be necessary. When an instruction word containing multiple instructions is transferred to this area, the buffer entry is freed. Instructions are transferred one at a time to the IR until the word is depleted of instructions.

## 2.1.  Basic Buffer Fill Technique

Various schemes for buffer fill exist that differ in detail from our example. One scheme is to divide the buffer into two portions, an upper and a lower half. The upper half is filled with instructions when the address of the first instruction to be fetched is computed. When the half buffer is filled, the processor is permitted to start instruction interpretation and execution. During this time the lower half of the buffer is filled with instructions. The hope is that the flow will be uninterrupted by a branch so that the address of the lower half of instructions can be easily computed when the address of the first instruction of the upper half is known.

The expectation is that when the processor reaches the boundary the lower half will be filled and the processor may proceed while the upper half is being refilled.

Consider a machine with a decode stage of 2 cycles, a place to put decoded instructions, and an instruction buffer containing eight instructions divided into two four-instruction half buffers. Assume no interference for memory reference or other delays in execution. Every 2 machine cycles the I-unit will require that another instruction be available for preparation. It will take 8 machine cycles to transverse a half buffer of instructions. If the memory speed to processor speed ratio of this system is 8 to 1, and if the machine has a four-way memory interleave, then the memory should be able to load four words of instructions in those 8 cycles. The I-unit will characteristically not have to wait for the delivery of instructions in a branch free instruction stream.

## 2.2.   Flexible Priority Driven Buffer Fill

There are more elaborate schemes for controlling the contents of an instruction buffer. In a system that has buffered memory interface registers, it is possible to change the priority of operand storage and fetch requests relative to the priority of instruction fetch requests. A set of pointers indicates the instruction currently in I-time and the instruction last delivered. When the I-time pointer gets too close to the last transferred instruction, the priority of instruction fetch is made higher than the priority for operand fetch or store. The execution of the machine will slow since it cannot get to memory to drain or fill registers. All memory cycles are given over to instruction fetch. When the processor falls safely behind the priorities are switched again. The design assumes an interleaved memory where contention between operand and operator memory references is very high.

## 2.3.   Short Loop Mode

Another form of buffer contents control is associated with branching instructions. A buffer can be put into *short loop mode* when the instructions in the buffer form a short loop such that a branch instruction repeatedly transfers to an instruction in the buffer. The condition may be realized by hardware detection or by the issuing of an instruction that places the processor in short loop mode. When in short loop mode, instruction buffer fill is suppressed until the branch leaves the loop or an exit from short loop mode is issued. To gain performance it is necessary that the time lost in refilling the buffer after leaving the loop is compensated for by faster execution of the loop. This means that a significant number of iterations must be expected and that the suppression of instruction fetch for the interval will speed the system by reducing memory contention delays.

Short loop mode effectiveness suggests somewhat larger buffers than would be required without the feature. The success of the feature depends on a high percentage of branches that end tight computational loops. Since such loops are often associated with vector or array processing, short loop mode addresses the same performance area as vector processing capability. The performance of a short loop feature will not match the performance of a vector processing feature since the vector processing feature not only eliminates many instructions but also reduces the number of branches that will be encountered in a program. The importance of reducing branches will be seen in the next section.

## 3.   BRANCH INSTRUCTIONS AND INSTRUCTION BUFFERING

This is a topic that involves memory as well as processor design; we now have enough information about both to investigate methods for improving the speed of a processor when it encounters branch instructions.

The instruction buffer works well for sequential instructions because it can predict well into the stream of instructions what instructions will be executed. A conditional branch instruction, however, breaks into the stream so that instructions brought into the instruction buffer may not be those executed by a processor.

Part of the solution to this problem involves the lookahead capability of the processor used with an alternate (smaller) buffer area. The I-unit moves forward in the instruction stream, preparing instructions and passing them off to E-units. When the I-unit encounters a branch it computes the address of the branch transfer address and requests that instructions from that address be sent to the alternate buffer.

When the branch instruction is executed the I-unit determines that instructions from the alternate buffer or the primary buffer be processed, depending on the direction of the branch. If the branch is taken, next instructions come from the alternate and additional instructions further down the branch path are brought to the primary buffer.

In order for this scheme to work, of course, it is necessary that the target address of the branch not be modified during the interval from its computation until the branch is actually executed. This means that the instruction itself, and the contents of index and base registers that have participated in the formation of the branch address, may not change.

This design reduces the time required to react to a branch by eliminating the need to reference memory for instructions to be executed after a branch. There is still a period during which the system may experience a delay, however. High-performance uniprocessors desire to maintain an uninterrupted flow of instructions from the I-unit. During the time required to determine whether the branch will be taken, it is not possible to know what instructions should be forwarded to the execution units. This period is called the *zone of uncertainty*.

Some machines attempt to reduce the delay caused by a branch by essentially trying to behave as if the branch instruction had not occurred. This takes some cooperation from the execution units. The I-unit makes a guess that the branch will not occur and prepares instructions along the primary path. It submits them to E-units, or to buffers between the I-unit and E-units, with an associated flag indicating uncertainty. When an execution unit encounters such a flagged instruction it does not execute it until it receives an OK signal from the I-unit. This allows the I-unit to prepare instructions beyond the branch and pass them off for as long as buffer space allows. If the branch is not taken, the E-units have a population of already prepared instructions for execution.

Some very aggressive designs for handling branch instructions have been proposed in the past. It is possible, for example, not only to prepare instructions on both paths, but to actually execute instructions in multiple paths. When the direction of the branch is known, arithmetic registers that hold values appropriate to the paths taken are used and the contents of registers used for other paths are discarded.

Another aspect of high-performance design is to try to improve the guess that the I-unit makes about which path the branch will take. Programming conventions frequently recommend that a program be developed in such a way that the path most probably taken is the fall through (no branch) condition. However, such programming conventions are, at best, uncertain.

Short loop mode indicated that a processor can be aware of the direction of a branch. A generalization of this idea involves the provision of a hardware *branch history table* that represents some set of recent branches with an indication of whether the branch has occurred. When the I-unit encounters a branch it determines whether it is in the branch history table and assumes that the branch will go the same way it went the last time it was encountered. The depth of the branch history table provides a predictive zone for the processor so that it can make better guesses about how a branch will behave.

Part of the art of branch history table design is maximizing the effective predictive zone by minimizing the number of branches actually recorded in the table. One approach is to record only branches that have been taken on the assumption that they will be taken again. Some distinction can be made between types of branches. Statistics may show that the probability of BRANCH ON INDEX instructions making the branch is much higher than for branches depending on data values. Most effective use of branch history table space is made by recording only those branches about which good statistical guesses cannot be made.

### 4. OPERAND BUFFERS

A simple form of operand buffering exists in any design where there is a super-architectural transfer of data. Operand values may be contained within the transferred group of locations. These operands may then be transferred from the memory interface buffer area to working operand areas on reference. This limited lookahead is possible with any broad interleave design.

A fundamental difference between instruction buffering and operand buffering is that the locality of operand reference is significantly lower than the locality of reference for instructions. Therefore to be effective operand buffers must be large or based on some predictive mechanisms for operand use.

We have already seen some forms of operand buffering and operand lookahead. Operand forwarding from the I-box is one form, and stack organization also gives some properties of operand lookahead. Small operand buffering areas requires that a predication be made as early as possible so that an operand request can be made and the operand arrive before it is needed by an execution unit.

The determination of needed operands can be represented in the architecture by the presence of instructions that ask for the transfer of logical units of data into a *scratchpad memory*. The address space for this scratchpad may be unique or may be some portion of address space that is mapped into a faster memory technology. The program takes the responsibility for determining what units of data are desired in the faster memory and issues instructions to transfer that data. Units like records or portions of arrays are reasonable examples of logical units that have a strong locality of reference for a period of time.

## 5. CACHE

A design of major interest to support both operand and operator buffering is an implicit local storage called a *hidden buffer* or *cache*. Generally a cache is a combined instruction and operand buffer that serves the I-unit, and all execution units, as general memory that is speed matched with the processor logic. The contents of the cache are managed by a hardware algorithm which attempts to maximize the probability that a particular memory reference can be satisfied with the contents of the cache. Cache sizes vary widely, ranging from very small areas in small processors to 64K bytes in very large processors with basic machine cycles in the 25 to 50-nanosecond range. The required size of a cache depends on the architecture of the machine, the instruction retirement rate, and the processor speed to memory speed ratio. Limited lookahead is applied to cache references so that the cache reference time is speed matched with the processor. This lookahead has the intent of enabling a cache reference to occur on 1 apparent cycle. There are actually 2 cycles involved in a cache reference, a cycle to determine if the referenced address is in the cache and a cycle to bring the contents to other registers of the processor.

Cache designs differ in the organization of storage, the units of transfer, the methods for determining available cache space, the instances when writes from cache to memory will occur, and other details. All caches, however, share the common goal of attempting to run the system at some close approximation of processor speeds. A goal of cache design is to achieve *cache hits*, finding referenced instructions or data in the cache, well above 95% of the time.

Cache may be a significant design attribute of both large and mid-sized machines. Some mid-sized machines, for example, the IBM 4341, get very good price/performance ratios with the use of cache because:

1. The design can use slower memory and simplified memory organization. Cache fill is accomplished by wider data paths from memory rather than by interleave designs. Thus cache can overcome the interleave granularity problem of mid-sized machines.

2. Speed can be achieved with simpler processor organization and lookahead. Instruction buffering units are not necessary. Reasonable lookahead comes from the cache design. Because super-architectural transfers to the cache are implicit in cache design, the probability of a successor instruction's being in a cache are high. Larger machines may use instruction buffers besides the cache. In such machines transfer to instruction

buffers is from the cache or additional buffers are only used on the DOR side for I-box output.

3. The amount of cache memory required for effective use of fast memory is significantly less than unmanaged scratchpads to achieve the same processor MIPS rate.

To demonstrate the importance of cache in the performance of a processor, let us consider a processor that completes an instruction in an average of 4 machine cycles with a memory that requires 32 machine cycles to satisfy a reference. This is an 8 to 1 ratio. We factor in the detail that not all instructions will contain memory references so that the average number of memory references is 1.7 per instruction. The following table shows the performance of a processor with a 50-nanosecond machine cycle. The table is computed using an equation that estimates the MIPS rate of a processor as follows:

$$MIPS = \frac{1000}{CT(AC + AMR \times MD \times PM)}$$

where CT is cycle time, in this example, 50 nanoseconds; AC is average number of cycles, in this example, 4; AMR is average memory references per instruction—we use 1.7; MD is the number of additional cycles required when something is not in the cache, in this example, 30; and PM is the percentage of time that we must go to memory.

$$MIPS = \frac{1000}{50(4 + 51 \times PM)}$$

where PM= .00, .01, .02, .03, . . . , .10.

| Hit Ratio (%) | Effective MIPS |
|---|---|
| 100 | 5.0 |
| 99 | 4.4 |
| 98 | 3.9 |
| 97 | 3.6 |
| 96 | 3.3 |
| 95 | 3.1 |
| 90 | 2.2 |

The importance of high cache hit ratios can be seen clearly. A 1% decrease in cache finds causes nearly a 10% decrease in effective processor instruction execution rates.

There is a trade-off between making caches large enough to ensure very high cache hit ratios and making the primary memory faster so that less of a penalty is paid for having to go to memory. The following table, computed in the same way, assumes only a 14-cycle delay for a memory reference. The speed of memory has been doubled to achieve a ratio of 4 to 1. Sixteen cycles are required for a response from memory, of which 14 contribute sequential delay.

$$MIPS= \frac{1000}{50(4+23.8\times PM)}$$

where PM= .00, .01, . . . , .10.

| Hit Ratio (%) | Effective MIPS |
|---------------|----------------|
| 100 | 5.0 |
| 99 | 4.7 |
| 98 | 4.5 |
| 97 | 4.3 |
| 96 | 4.1 |
| 95 | 3.9 |
| 90 | 3.1 |

This processor responds less dramatically to cache misses and can tolerate a 3% ratio and still perform at the level the first processor can perform at with a 1% ratio. Therefore it can tolerate a cache that is smaller than the other and will be effective only 97% of the time.

The question that faces a designer is which of these machines is a better price/performer in a variety of environments with different operating systems, multiprogramming levels, and so on in a particular technology period where the costs of more cache and control for the cache is traded off against the cost of faster primary memory.

### 5.1.  Cache Mapping and Loading

The description of cache design we will use is derived from features of caches used in the IBM 3033 model of the 370 generic product line.

Figure 27 is a picture of the organization of IBM 3033 cache. It shows a cache, a cache address unit, and a memory. The cache address unit represents a mapping of the memory locations that are currently in the cache. In the largest of the caches used in IBM 3033 design there are up to 64K bytes in a cache unit. Other caches have 32K, 16K, 8K, 4K, and 512 bytes for support of this architecture.

A 64-byte cache contains:

1.  *Blocks of Eight Doublewords.*   Each doubleword contains 8 bytes. Each *block* contains 64 bytes. A block is the logical unit of interaction between memory and the cache.

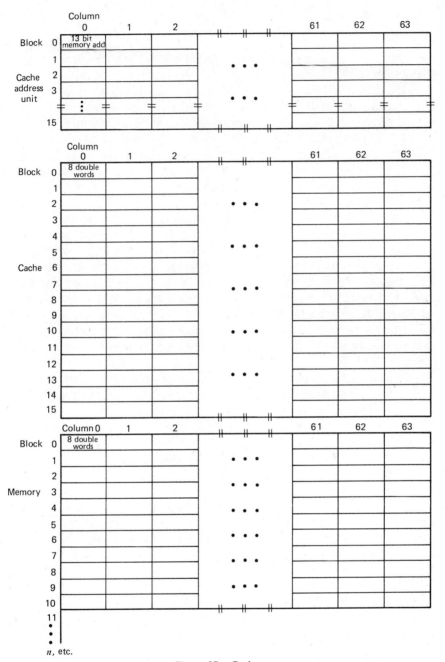

**Figure 27.** Cache.

2. *Columns of Blocks.* Each *column* contains 16 blocks. Each column of
cache therefore contains 128 doublewords, 1024 bytes.

In a 64K cache there are 64 columns. Total provided cache storage therefore
contains 1024 blocks, containing 8192 (8K) doublewords, 64K bytes.

In the IBM 3033 the granularity of a memory unit is 8-byte doublewords. A
fully configured IBM 3033 memory is eight-way interleaved. A block of memory contains eight sets of doublewords, each memory unit contributing one
doubleword to the block. A memory block therefore contains 64 doublewords.
The addresses contained in block 0 are 0 to 63. The addresses contained in
block 1 are 63 to 127, and so on. Blocks in memory are also organized into 64
columns. Since memory size is greater than cache size, each column of memory
contains more blocks than a column of cache.

When transfer is accomplished between memory and cache, the unit of transfer is the 64-byte (eight-doubleword) block; one doubleword is contributed
from each memory unit. This 64-byte structure occupies a block of the cache.

A cache block load is performed when data not in the cache are referenced
for a LOAD or a functional instruction requiring data from memory. Eight
doublewords are transferred. The memory unit that contains the address of the
referenced data within its doublewords delivers to the cache and to registers
within the I-box first. The other seven doublewords are subsequently delivered
to the cache block. The goal is to deliver the additional doublewords one every
machine cycle.

The address unit associated with the cache is organized by *block address
registers* within columns. Each block address register corresponds to a cache
block. A block address register contains the address of the memory block
currently residing in the cache block that corresponds to the block address
register. A block address register also contains bits to indicate that the data in
the cache is valid and usable. A column of block address registers corresponds
to a column of the cache.

The address first formed by the system is a virtual address. It is formed by
use of the address field in the instruction and base and index registers. This
address must be converted to a real address by use of an address translation
mechanism (TLB) of the type described in Chapter 10. When the real address is
formed, a search of the cache address array is made to determine if the referenced data is in the cache.

In IBM bit numbering conventions, bits are numbered from 0 on the left
upward to the right. Address bits are interpreted in IBM 3033 24-bit addressing
form (assuming a 64K buffer) as follows:

| Bits | Use |
|---|---|
| 21–23 | Byte within doubleword |
| 18–20 | Doubleword within block |
| 12–17 | Address unit column |
| 0–12 | High-order memory address |

Each address unit block address register contains a 13-bit, high-order memory address. This address indicates the specific area in memory from which the block in the cache has been taken. The high-order memory address bits from the real address are compared against the contents of the cache address unit.

The address unit column bits determine which column of the cache address unit is to be searched for the high-order address. The 6 bits allow for addressing the 64 columns of the cache address unit. The indicated column block address registers are searched to determine whether the address indicated in a block entry corresponds to the address being used. This search is an n-way parallel search that enables the determination of presence to be made in 1 cycle. N for the 64K cache is 16.

The above design suggests that any block of memory cannot go to any block of the cache. Any block in a memory column can go only to any of the 16 blocks in the corresponding cache column. Thus there is a partitioned mapping of memory to cache. We refer to the partitioning again further on.

If a search in the address unit is successful, no main storage reference need be made. If the search is unsuccessful there will be a main storage reference. This main storage reference is distinct from a paging operation that is undertaken when virtual memory translation fails to find a page.

## 5.2. Cache Contents Management

If a cache load is undertaken then space must be found in the cache for the new data coming from memory. A form of least recently used algorithm determines which of the blocks in the cache is to be overlayed by a new block. This overlay is required if no block contents control bits indicate an empty block. In a column partitioned mapping, such as we have been describing, the search is for least recently used block within a column. This search is supported by a *replacement array*. A replacement array shows the order of references to blocks in a column. Each time a block is referenced by an activity that takes data from the cache, the replacement array is modified to show that activity. For each block in the column there is a replacement array value. The set of replacement array values provides an ordering of blocks by recency of reference.

There are some rather simple and elegant schemes for mapping recency of use in the cache. One technique is to represent the blocks in a small fast memory in matrix form. Whenever a reference to a block is made a 1 is written across the row representing the block and a 0 is written down the column representing the block. This technique will cause the numeric value of any row to represent an order of recency of reference.

When a cache block is selected for a new memory block, the new memory block is brought in in the manner described above. The first doubleword is the one containing the actual address, and subsequent doublewords are delivered in some order, perhaps at the rate of one per processor cycle. The delivery of the actually addressed block enables a processor to start work before the completion of the entire block load.

### 5.3.  Stores and Changes to Cache

When a reference to memory is a store reference and the address is found in the cache then both the main memory and the cache entry may be changed. A cache that does an immediate change to main memory is called a *store through cache*. The cache in the IBM 3033 is a store through cache. A storage reference to data not in the cache records the store directly into memory. It does not cause data to be brought to the cache, nor does it affect the replacement array.

Input/output operations in main memory do not flow through the cache. For an output operation, data flow from main storage directly to the output paths. For an input operation, a check is made to see whether addresses associated with the input area are in the cache. If they are, then an invalid bit is set in the address mapping unit and data flow into main memory. The next reference to cache will find the invalid bit indicating that the cache does not contain the current data in the represented memory locations.

### 5.4.  Some Concluding Comments

There are many variations in cache design. Specific designs will reflect processor and memory organizations. Variations include the size of cache, the size of the transferred unit, the details of least recently used algorithms, and so on. A rather complex analysis of addressing patterns usual in the architecture is necessary to determine proper cache size, organization, and dynamics.

The cache described above is partitioned so that only blocks of a specific memory column can go into the blocks of a specific cache column. This is typical of larger caches. The partitioning is done so that the addresses held in a block address register can be smaller and so that the extent of the address unit search can be limited. Associative memories with parallel search are expensive. It is important for cost control to limit the extent of the parallel search that must be made for a cache address determination. The above example limited the search to 1/64th of the address registers by limiting the search to one column. The search must, of course, be fast and parallel so that time required to find an address is minimum.

Smaller caches are frequently not partitioned so that any location in memory can go to any area in the cache. Theoretically, partitioning of buffers leads to less efficiency than no partitioning. This is true of any general buffering system. Smaller caches can avoid partitioning because address units may be small enough to be completely searched. Partitioned larger buffers, however, seem preferable on larger machines. Partitioning will affect the least recently used algorithms as well as the memory-to-cache mappings. Nonpartitioned caches may use a universal, least recently used algorithm to find the least recently used block in the entire cache rather than in the column.

There are some problems with cache contents management in systems that are intensively multiprogrammed or that rapidly move back and forth between supervisor and problem state. After return from the operating system a nonpartitioned cache will consider that the operating system blocks are most recently used, and, when additional cache area is needed, will overwrite data blocks of an application program that are still current but are not the most recently used because of the operating system interval. For reasons of this type there is currently interest in the notions of specialized caches, instruction caches, operand caches, supervisor state caches, and so on.

There is an interesting relationship between operating systems design, general memory allocation issues, and the organization of cache. Although theoretically a nonpartitioned cache is more efficient than a partitioned cache, the partitioning can produce the same effects as special-purpose caches when allocation of memory is carefully controlled. Thus if the operating system only exists in certain columns it cannot displace application code in other columns.

Some caches are not store through caches. When a change is made to the cache, a bit in the block address register is set to indicate that the contents of cache are not reflected in memory. When the block loses its place in the cache the changes are stored in memory. Caches may also communicate directly with the I/O system. In such a design, used in some non-IBM models of the generic architecture, input flows directly into the cache.

There is some interest in providing specific cache control instructions. Incremental improvements in cache hit ratios that might effect significant improvement in sustained MIPS rates might be achieved by issuing instructions that can control cache contents with more refinement than least recently used algorithms. This moves the concept of cache, which is a design and not an architectural notion, into the architecture of the system.

Finally there are implications for program organization and design implicit in the cache. Some data structuring techniques, for example, the use of linked lists to associate data structures, are not efficient in cache. Simple linear tables of addresses would enable searches to be made with fewer cache loads than sparse linked lists spread across multiple blocks would.

## 6. OTHER HIERARCHICAL NOTIONS

It is possible to conceive of two (or more) levels of memory without the concept of a cache. For example, some systems have the notion of two levels of memory, a faster, smaller memory close to the processor, and a slower, larger memory behind. This structure may be exposed to the architecture so that a program is responsible for moving data from one level to another. This explicit movement by program may exist at a level of the hierarchy other than the cache/main-storage level. Instructions for the transfer of blocks of data or instructions are executed in a way that makes the back memory look roughly similar to an I/O device. In addition, and this is common to many such hierarchies, the processor cannot execute instructions from the back store, and the granularity of reference is limited to large blocks of transfer. The only instructions that can execute on the back memory are GET and PUT instructions that move data between faster and slower memory elements.

There is a growing interest, actually a reactivation of interest, in providing memories behind the main memory. The initial reason for such memories was to provide space for *roll-in* and *roll-out*, the movement of code and data, under operating system control, to a backing store when they are not being used. Similarly, I/O devices were staged into the backing memories, and large buffer areas could be defined in these areas.

The contemporary use of such boxes of large memory with relatively slow (microsecond) access time is not very different. They are intended for use as paging areas, replacing drums and disks as the units where a paging sytem holds its moved page frames, and as possible data buffer areas for use by data management software.

Some designers look forward to certain systems extensions where memories will be used as an interconnection mechanism between processors that exist at some distance from each other. In a multiprocessor design, processors share memory under the control of a single operating system. In some new design concepts processors will have their own population of devices, a significant private memory, and private local operating systems but use a shared systems memory as an area for sending messages that are requests for data to be transferred from one system to another. Similarly, some conjectural designs call for the ability to page from storage devices either directly into the memory of the system to which the paging devices are attached or remotely over fiber optic lines to the memories of remote computing systems.

## 7. INSTRUCTION AND CONTROL MEMORIES

It is fundamental to the idea of the Von Neuman machine, the traditional organization of a computer, that instruction and data be intermixed in a single memory. The concepts of instruction and operand buffering dilute this notion to some extent, but even a machine with these staging memory areas and with a cache, preserves the concept that at the first architecturally defined level of memory instructions and data are intermixed.

Rather early in the development of computer systems, however, some designers had the idea that instructions should be placed in memories separate from the memories that held data.

The concepts behind the separation of data and instructions are complex and come from programming structure concepts, security concepts, and performance concepts. In early systems that had instruction and data memories the memories were essentially the same in physical implementation and very similar in speed. The instruction memories would not be modifiable during execution time and would be able to write only under special circumstances, perhaps at a time when a special key was inserted into the system.

An extension to the idea that instructions may be placed in a special memory is the concept that the architecture of the machine itself may be represented in memory. Memory has become not only a vehicle for data and programs but for the implementation of the logic of a system. The notion of a *read only memory* (*ROM*) is well established, as is the notion of a *programmable read only memory* (*PROM*) and similar memories that can be written only under special circumstances. These memories are frequently used to define an address space that is beyond the direct reach of instructions in the architecture and are used to contain instructions that are executed by small processors to simulate the architecture of a higher-level system. These small control memories may contain, in addition to certain features of the architecture, certain programs of the operating system. An example of the use of control memory for support of the architecture is the implementation of a MULTIPLY instruction by a program in a control memory executed by a small processor whenever the MULTIPLY instruction is issued. An example of the use of control memories to support operating systems is the placement of certain interrupt handling code in the operating system in a control memory. In general, implementation of functions in control memories rather than hardwiring slows a machine and implementation of operating systems functions in control memory speeds a machine. A fuller discussion of these issues is undertaken in Chapter 21.

## 8. FINAL COMMENTS ON MEMORY

One of the truly dramatic changes in computer design and implementation is the way memory is packaged and powered. Ferrite core memories would occupy large amounts of space, while today a 64K-byte memory of 150 nanoseconds access time can be placed on a hand held chip. The power requirements have been seriously reduced and technology exists for powering only those locations of memory that are currently being referenced. The availability of large, fast inexpensive memories has more than increased the memory sizes of processing units; it has, combined with advances in processor logic, transformed the way machines may be built.

It is important to observe that we are now in a technological era where the traditional speed imbalance between memory and processors can be addressed in a very rich variety of ways. Memories of significant size are becoming fast enough and cheap enough so that processors of impressive performance capability that are not faster than their memories can be developed.

It is possible that in the future memories of speeds from 25 to 500 nanoseconds will be available in large sizes. What is not certain is what the price differences between these memories will be or what price differences in processor speeds will be across a broad range of performances.

It is possible that in the future processors that achieve greater than 2 or 3 MIPS can have speed matched memories. This will simplify design by eliminating the need for caches. The very high-performance machines of the near future, however, will still face the memory imbalance problem. Designers of the future will have a series of choices to make about processors, the variety of memories used, the speed balances between memories and processors, and the development of hierarchies in order to achieve price/performance goals.

## QUESTIONS

1. What information is needed to determine if an instruction about to enter I time is in the buffer?

2. Why may instruction lookahead be suspended?

3. What problem does conditional branching present to lookahead?

4. What is the purpose of an alternate buffer area?

5. What is the purpose of a branch history table?

6. What is the importance of cache?

7. What is a cache address array?

8. What is a store through cache?

# Chapter 20

# Input/Output
# Design

## 1. BASIC CONCEPTS

This chapter discusses aspects of the design of the I/O flow of a computer system. The design of I/O involves several decisions concerning:

1. The devices that can attach to a processing system.
2. The interconnections between I/O devices and the processor/memory.
3. The ways that I/O operations are coordinated with other activities in the processor and memory of the system.

The design of devices that attach to computer systems is a highly specialized world. We address ourselves primarily to the interconnections and the distribution of functions across the paths that are used to interconnect.

Chapter 11 introduced some concepts of the architecture of I/O operations. We saw that a complete I/O command must designate:

1. The function to be performed.
2. The device it is performed on.
3. For some devices, a location in the device where the data is to be stored or found.
4. A pathway to the device.
5. The set of memory locations from which or to which data are to be taken.
6. If there should be an interrupt when the data transfer is finished, when the device is free, or when the I/O operation cannot be performed for some reason.

The architecture of I/O varies in the method used to supply this information, how much of the information may be implied, and how much of the total I/O function is accomplished in applications programs as opposed to the operating system.

The design of I/O to support an architecture may vary as to:

1.  The distribution of control activities along the pathway from processor to device.
2.  The number of separate I/O pathways connecting devices to processor/memory.
3.  The flexibility of attachment of various devices to various paths.
4.  The degree to which I/O may be overlapped with processor activity or other I/O.

We begin by describing the simplest form of I/O interconnection and then become more and more elaborate until the design reflects the attributes of contemporary larger systems.

## 2.  PROGRAMMED INPUT/OUTPUT

Figure 28 shows a tape device that is attached to a processor by a pathway that is 16 bits wide. We assume, for simplicity, that the unit of transfer between the I/O device and the processor is a 16-bit word and that the processor word size is also 16 bits. The 16-bit-wide path to the tape unit is used for both control information and data that are to be read or written on the tape.

The tape unit has sufficient logic to enable it to react to commands to write, read, rewind, reverse direction, find a starting point (a tape load point position), and send status data. The tape can read or write various size *blocks*. A block is a unit of information that is separated from other units of information by a space between blocks.

The tape unit has a 16-bit buffer that is used, for input functions, to hold data that are being written one character at a time and, for output, to hold data that are being read one character at a time. Data are recorded on tape in 9-bit frames, each one of which moves sequentially under a read-write head. One of the bits is a *parity bit* that is set to 0 if the sum of the bits in the other eight frames is even and to 1 if the sum is odd. This is used to check for the correct reading of data from the tape to the tape unit buffer. The unit of transfer between tape and processor is 16 bits, two tape frames making one word, stripped of parity.

In order to read or write, the processor must address the desired unit. This address actually has two potential components, the address of the bus that connects the unit and the address of the unit itself. Since in this example Figure 28 shows one device per bus, the two-element nature of the address is not important to us yet.

The processor must also provide the function to be performed, the location of data to be transferred, and perhaps other control information for the device.

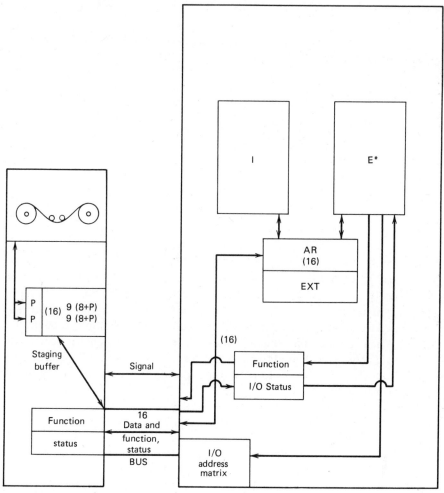

**Figure 28.** Programmed I/O. *E-unit performs I/O, selects unit, sends function and control codes, tests status, and sends/receives data in AR.

For this example we assume an I/O instruction of the following form:

OP CODE: Bits 0–2. (Always 011 to signify I/O.)

REGISTER ADDRESS: Bits 3–4. (Specifies register from which data is to be transferred to device or register that is to receive data from device.)

FUNCTION: Bits 5–7. (Specifies input, output, no transfer, status test, etc.)

CONTROL: Bits 8–9. (Specifies device start, device stop.)

DEVICE ADDRESS: Bits 10–15. (Specifies address of device to be controlled by program I/O.)

Figure 28 brings the I/O data flow directly to the arithmetic registers of the processor. This characteristic of data flow is called *programmed I/O*. Arithmetic registers are the source and destination for data. Thus before doing input the register must be made available. Before doing output a value must be placed in an arithmetic register.

The transfer of one word to the tape unit might look like the following:

> START TAPE1, BUS1, NO TRANSFER
> LOAD R1, A
> WRITE TAPE1, BUS1, R1
> STOP TAPE1, BUS1, NO TRANSFER

Three separate instructions are shown, START, WRITE, and STOP. This may only be an assembly language convention. The three instructions may have a single I/O instruction format with variations in the function and control bits.

The START instruction begins the tape movement across the space between blocks. When the E-unit recognizes the START instruction, it uses a device addressing network to address the device and sends function and control codes to a processor I/O function register. From this register they are transferred to the addressed device. When the device accepts the transfer of function and control codes, it causes the start of the function. In this example, tape motion starts. When the tape movement is up to speed, a done signal is sent to the processor. This signal is used to move the next instruction into E-time. For the duration of the time it takes to reach tape speed, the processor is hung on the START instruction.

After the arithmetic register is loaded with the vaue to be written, another I/O instruction is issued. This WRITE instruction causes the function and control bits to be sent to the addressed tape unit and the contents of the designated arithmetic register to be sent to the buffer in the device. The processor passes on from the WRITE instruction when the tape unit sends a done signal. When this signal is received the processor goes on to issue the STOP instruction, waiting to proceed until a done signal is received from the tape unit.

All of the above I/O instructions have available to them a set of device status indicators for each device. These status indicators represent states such as device busy, device done, device ready, device malfunctioning, and other indicators that are device specific. A precondition for starting an I/O is that the indicators show a proper state for the beginning of the function. Control and function coding in the instruction may determine the status required for the operation to proceed.

Characteristically the data on a tape unit is written in blocks of multiple words or characters. In order to accomplish this the following loop may be written to write 100 characters on tape:

> LOAD INDEX1, "50"
> LOAD INDEX2, "0"

```
                   START TAPE1, BUS1, NO TRANSFER
        WRITE: LOAD AR1, A(I), INDEX 2
               WRITE TAPE1, BUS1, AR1
               INDINC, INDEX2, "1"
               INDECT, INDEX1, WRITE
               STOP TAPE1, BUS1, NO TRANSFER
```

The " " signs around the values in the two index register load instructions and the index increment instruction signify that these are immediate instructions with the constants to be used imbedded in the instruction.

The above coding initially loads two index registers with values that will control the movement of data. Index register 1 counts down for a 50-word transfer, index register 2 participates in the formation of the address for consecutive words to be transferred to the tape unit. Each time a word is written, index register 1 will be decreased by 1 and tested for 0. The INDECT instruction causes a transfer back to WRITE if the index register is not 0 and falls through to the STOP instruction when the index register is 0. After each word is written, index register 2 is increased by 1 in order to advance to the next address. Words written on tape are A(1), A(2), A(3), . . . , A(50). Since each word contains two characters, 100 characters are sent to the tape.

The fundamental feature of this programmed I/O is the flow of data through the arithmetic registers and the detention of the processor through the activity of the I/O instruction. This means that there is no effective overlap between the processor and the I/O devices. This is true for both input and output. A loop to cause the reading of 100 characters (50 words) of data might look like:

```
               LOAD INDEX1, "50"
               LOAD INDEX2, "0"
               START TAPE1, BUS1, NO TRANSFER
        READ: READ TAPE1, BUS1, R1
               STORE R1, A(I), INDEX2
               INDINC INDEX2, "1"
               INDECT INDEX1, READ
               STOP TAPE1, BUS1, NO TRANSFER
```

This loop stores consecutive tape loads of AR1 into consecutive memory locations beginning at A(1) and ending at A(50).

In the example so far each READ or WRITE loop assumes that the device is ready for the function or will soon become ready for the function. A READ or WRITE that finds a buffer that is not ready to transmit or receive data merely waits until the proper signal is set. There is no need for coordinating instructions of any sort because the processor is wholly devoted to I/O until the I/O function is completed. Certain malfunctions of the unit, of course, might hold the processor at an I/O instruction indefinitely. We discuss this a bit further along.

Each READ or WRITE involves only one word of a multiword block. Correct operation requires that each READ or WRITE must be issued within a fixed interval. This interval is determined by the characteristics of the device. The relevant characteristics are device motion speeds, space between recording areas, and data transfer time. These considerations apply to any tape or disk that has a concept of blocked input and output grouping individual characters or words. There are devices that do not have this concept. A single READ or WRITE constitutes an entire I/O operation on such devices.

Before leaving the basic I/O example, let us consider the problem of malfunction. To handle device malfunction some protective coding might be necessary. Some devices have an automatic retry feature that is used if a malfunction occurs during an operation. On a tape, for example, this automatic retry may backspace the tape and try to write or read the current frame again. Other tape units put down bad spot detection patterns when a faulty write occurs and try to write correctly after the bad spot pattern. When reading, the unit recognizes and bypasses bad spot patterns.

When retry fails, however, the design we have described so far merely stops the processor and probably shows some malfunction indication on the console of the system. It is then the responsibility of the operator or maintenance person to confront the error.

A way of testing the status of a device can be provided by making the status indicators available to a TEST I/O instruction. Such an instruction does not cause any activity but merely skips or branches when a device malfunction or a device busy condition is determined.

An I/O instruction may be preceded or followed by a TEST I/O instruction. When a functional I/O instruction experiences a malfunction it is terminated, that is, released from E status. The signals may then be tested by a TEST I/O instruction that tries to determine the cause of error and then enters some coding that will undertake a recovery procedure indicated by program code.

This example has used a magnetic tape unit. Other devices may be used with programmed I/O. The I/O commands may be the same, but it is common that each device take a somewhat different sequence of commands and that the interpretation of the function and control bits may be somewhat specific to each device. For example, a paper tape reader that does not have the property of continuous motion between reads or writes may not need a STOP or a separate START with no transfer but may be addressed only by READ and WRITE that cause movement and recording or reading. A disk unit may require a separate SEEK instruction to locate a record on disk surface. A different interpretation of the status bits may also be made as a result of the device addressed.

## 3.  PROCESSOR OVERLAP

Let us now consider some timing considerations in I/O more closely. As we have described the START, WRITE, and STOP instructions, the processor is considered busy until a done signal is sent from the device. In the above discussion the done signal was sent only when the tape was at full speed (START), when the data had actually been written on tape (WRITE), and when the tape had actually stopped (STOP). This means that the processor could do nothing during the entire I/O operation but wait for the done signal.

It is possible to have the I/O device send signals at different times so that some processor instructions might be executed while certain I/O functions are being performed. Exactly when a signal may be sent depends on the function and the design of a device. For example, a tape unit may send a signal for a START any time after it has received the START directive until the tape is at full speed. If it sends a START undertaken signal there is an interval of time during which the tape is running that may be used by the processor to execute processor instructions. In such a design the START instruction is moved to the I/O device but is retained in E-time only until the START initiated signal is received. When this signal is received a successor instruction to the START may be executed by the processor. Thus while tape motion begins arithmetic may be undertaken.

If such overlap is permitted, the synchronization of processor activity with I/O activity may be achieved by various methods. One method is to provide an instruction that the program must issue from time to time to see if a tape at speed signal has been received. The status test instruction approach may lead to *time-dependent coding* that is very tricky. A status test delayed too long may, in some I/O situations, result in malfunction of the system and lost data.

The ability of a processor to overlap some functions with I/O depends not only on coordinating mechanisms but on device intelligence. In order to have overlap there must be enough control circuitry in the unit so that the attention of the processor is not required to monitor the entire I/O operation. That is, the device must have the ability to furnish its own control, based on interpretation of the function and control bits sent to it, from the time it receives a directive until the time it is finished. In effect, some kind of specialized E-unit circuitry must be placed in the device so that the E-unit in the processor need not be busy with the entire operation.

It may be that a device has control over its own mechanical functions—it can start a tape, rewind, stop a tape, start an arm moving, and so on—but the E-unit and arithmetic registers are involved in any actual data flow moving between the processor and the device. The instructions that use circuits in the E-unit and/or registers devoted to I/O may not be overlapped with data transfer if this is the case. However, instructions may be executed while the device is getting up to speed and while it is stopping.

On a system that has devices that are sufficiently intelligent to control their own mechanical functions, overlap still depends on code sequencing. For example, the issue of a WRITE instruction immediately following a START may cause the processor to wait on the WRITE until the START done signal is sent. A program desiring overlap must either time very carefully or issue a status test instruction to determine when the WRITE can be issued without eliminating overlap. Alternatively, a skip subfunction may be associated with I/O instructions in the architecture and design. The processor would skip out of a WRITE instruction issued too early.

Such programming is very risky. There are time constraints on two sides. I/O instructions must not be issued too early to inhibit overlap, and they must be issued within a time interval. A block write to tape, for example, must occur within a certain period of time after the tape is brought to speed.

## 4.  MORE ELABORATE INPUT/OUTPUT DESIGN

Enhancements to architecture and design may provide a simplification of the coding patterns and an ability to execute some instructions while I/O data transfer is going on.

Reduced coding and timing requirements for overlap may be achieved by providing a place in the processor, other than the arithmetic registers, for data to flow through. Figure 29 shows a processor with I/O staging registers that hold data moving to and from memory without the use of the arithmetic registers. This design changes the nature of processor and I/O flow interference.

In addition to the data staging registers in the processor, Figure 29 elaborates the I/O interface registers with an address word register and a count register. We see the uses of these registers below. Figure 29 also shows a path between the E-box and the memory interface registers. The E-box uses this path to get access to memory for data involved in I/O operations.

In the processor of Figure 29, processor instructions and I/O instructions do interfere with each other at the memory interface registers. I/O transfers contend with LOAD and STORE instructions for pathways to the memory. Instructions that do not refer to memory can be performed. Resolution of contention for memory access between the I/O staging area and processor instructions is handled automatically. Since arithmetic registers are not part of the I/O flow they may be changed at will by the processor. Some additional constraint of overlap may exist if circuits in the E-unit used for instructions are needed to control I/O flow.

Over a particular interval an I/O staging register makes requests for data or for access to memory at a rate that may allow some memory cycles to be made available to processor instructions. Overlap between processor and I/O now includes instruction execution during device start and stop times, execution of non-memory referencing instructions, and execution of memory referencing instructions during intervals when I/O staging reference is not being made.

The design that allows this increased overlap implies a modification to the architecture. Since data does not flow through arithmetic registers, addresses in memory for data must be part of the I/O command sequence, not part of LOAD and STORE sequences. Some architectures use arithmetic registers to hold memory addresses, some allow memory addresses in I/O-oriented instructions. Whether the beginning address of a memory area to be used for I/O comes from an instruction or a register, it is placed in the I/O address register.

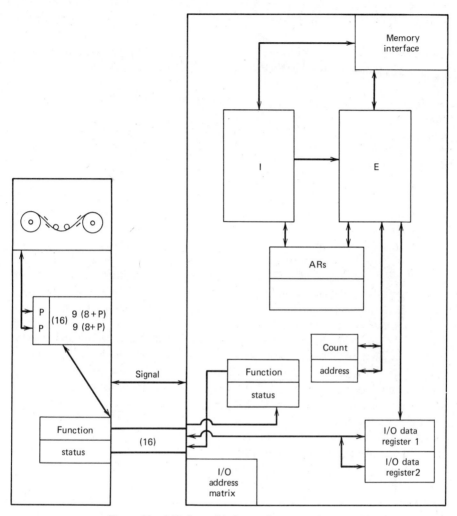

**Figure 29.**   I/O flow with data and count registers.

The I/O count register is used to allow a single I/O instruction to transfer multiple units of data. A count of addresses to be transferred is placed in this register and the I/O continues until the register is reduced to 0 by control circuits. This eliminates the need to write tight loops for multiple word I/O operations. Implicit in the design is the automatic adjustment of the I/O address and count registers whenever data are moved into or out of the processor. Each new data movement uses the new address to move a data value to or from the I/O interface. The I/O operation terminates when the count register goes to 0.

In this way a block of words may be delivered from the processor to an I/O device, or to the processor from an I/O device without a constant loading of registers. This permits the arithmetic registers to be freely used for computation during active I/O intervals. In addition, it frees the programmer or compiler from the task of timing I/O loops in order to deliver or receive words sufficiently quickly from a device, such as a tape or a disk, with some characteristic of continuous motion.

There is no danger now that data will not be delivered in time. However, the completion of the WRITE or READ must still be tested if an interrupt is neither available nor called for. In many designs the program has the option of calling for an interrupt at the time that the word count goes to 0 so that a status test instruction is not necessary.

This ability to transfer blocks of data to and from I/O should remind the reader of the I/O architecture that was described in Chapter 11. That architecture contained an SIO (START I/O) instruction that referenced a channel address word (CAW) that addressed a list of channel command words (CCW). An architecture of that richness can be supported by model dependent designs that determine, within price/performance constraints, exactly how much processor I/O overlap will be provided for on each model of the architecture. The design variations essentially determine whether I/O flow contention will be for processor circuits, for paths to memory, or merely for memory cycles so that no processor circuits are involved in the I/O flow.

There is an interesting detail in the Chapter 11 architecture of SIO, CAW, and CCW that we should investigate here. Notice that the CCW containing both initial transfer address and word count is contained in a memory location, as is the data to be transferred. In a design to support this architecture for a machine where minimum expense for overlapped I/O is desired, it is possible not only to take memory cycles for the delivery or fetch of I/O data, but also to augment the address and the decrement of the count. The transfer of data between the processor and the memory, therefore, requires three memory references, a reference for the data to be placed or fetched, a reference to obtain the address and count values, and a reference to place the adjusted values back into the memory. In addition to this use of memory, the processor will be delayed further if the data flow of the units transferred by I/O commands intersects with the data flow of processor instructions. The use of I/O address and count registers as shown in Figure 29 improves the overlap of this architecture by eliminating the references to memory to adjust addresses and counts. However, interference with processor instructions would still occur to the extent that processor arithmetic circuits and pathways were used to modify address and count values.

In calculating the speed of various processors, one part of any figure of merit is the *I/O interference*. Some vendors provide methods for calculating the impact on processor speed of various I/O patterns in terms of the degrees of interference with processor functions that an I/O event causes. The fundamental goal of design, as I/O designs become more ambitious, is to reduce the interference of I/O with processor operations.

## 5.  BUFFERED INPUT/OUTPUT

For devices that work with fixed units of transfer, such as 80 columns for a card unit or 128 characters for a print line, or for systems that have tape or disks with fixed block sizes, it is not necessary to provide a word count and it may not be necessary to provide an address for the I/O processor transfers.

Some systems have provided buffers between the fixed unit size devices and the processor so that all of the data that is to be transferred to or from the unit is collected in the buffer. Buffer fill is recognized by an interrupt or by the use of a buffer fill test or skip instruction. When the buffer is filled (on input) the processor executes a buffer to memory block transfer instruction that moves the contents of the buffer into the memory locations indicated. The processor can overlap the movement of data from the device to the buffer and is occupied only during the transfer from the buffer to memory. This transfer time occurs at memory rather than at device speeds so that the time that the processor is involved with I/O is reduced.

Some designs have provided for addressing directly into the buffer, which is associated with a set of fixed locations to or from which a device can transfer data. These addressable buffer locations are organized so that other memory references can be made while these buffers are filling or draining.

The intent of such designs, used very early in the history of the industry, is to free the processor while fixed units of information are transferred between memory and I/O devices.

## 6.  DIRECT MEMORY ACCESS ORGANIZATIONS

Figure 30 shows a system in which the path from an I/O device to memory does not go through the processor. Such designs are called *direct memory access* (*DMAC*) or *external memory interface* (*EMI*) designs.

The intent of the design is to minimize the interference of I/O with the execution of instructions in the processor. The processor is delayed only when an instruction fetch or an operand store or fetch is to be made from a memory bank that is the target or source of an I/O transfer. When contention arises, the processor always loses. This is because many I/O devices have a fixed interval of time within which they must gain access to memory because they have some characteristic of continuous motion. Tape and disk units fall into this category.

The pathway between I/O and memory is the *primary storage control unit* (*PSCU*) if one exists or the memory interface itself. I/O data flow and instruction/data flow contend beyond circuits in the processor. The degree of memory contention actually experienced depends on the speed and organization of devices and on the organization of memory partitioning.

A reasonable DMAC design provides for count and address fields to be sent to and maintained in an I/O device. The device is responsible for augmenting

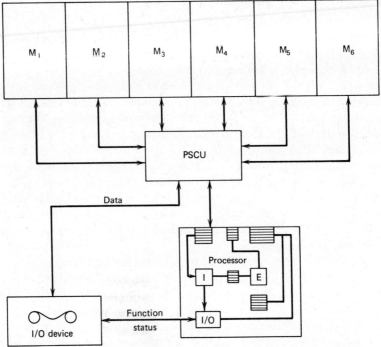

**Figure 30.** DMAC.

addresses for each data transfer delivered or taken, and for counting down the number of words (or characters) to be transferred. When the count is reduced to 0, the I/O device sends a signal to the processor denoting the end of data transfer. This signal may be used to generate an interrupt or it may be tested by an instruction.

We have dealt with degrees of overlap between a processor and a single I/O device. More elaborate designs attempt not only to increase the degree of processor I/O overlap but to introduce the possibility of multiple overlapped I/O operations. The next sections discuss environments with more elaborate I/O subsystem designs.

## 7. MULTIPLE UNITS

Previous sections have discussed the interconnection of a processor with a single device, a tape unit. Consider Figure 31. This figure shows a processor system with DMAC capability and a number of different device types. Interconnection between a device type is through a specialized bus or pathway (commonly called a *channel*) that takes on the name of the device type for which a connection is provided. Thus we speak of printer channels, tape channels, disk channels, and so on. The channels are specialized and matched to the data

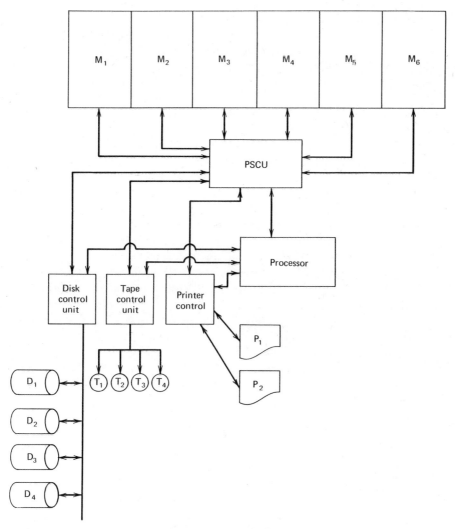

**Figure 31.** Multiple units.

transfer and signal characteristics of the device to which they provide a pathway. Channels may connect to *control units*, which provide connections to multiple devices of the same type. The control unit contains logic to control the operations of its devices so that all control circuitry need not be replicated in each device. Figure 31 shows a tape control unit attached to four tape units, a disk control unit attached to four disk units, a printer control unit attached to two printers, and so on.

A characteristic feature of control units for tapes and disks is that only one device can be actively transmitting or receiving data at one same time. The control unit provides a place where control functions can be performed. For example, the address and count values may be sent to the control unit. This allows for multiple control units to be transmitting or receiving data in parallel and introduces the possibility of read/write/compute overlap or various forms of multiple read/multiple write overlap with computing in the processor. The architecture need not change to reflect the fact that CCW-like information is sent to a control unit rather than maintained in memory or the processor. The interpretation of the I/O command changes in that the memory address given as holding the address and count fields is transferred to the control unit rather than maintained and updated in memory or in special registers in the processor. Thus all processor pathways and circuits are always free to perform non-I/O instructions. Contention between processor and I/O occurs only when they require access to the memory. The path between I/O and memory is the PSCU or its equivalent, and the processor does not lie on the path.

If multiple devices are to be run in parallel, however, some provision must be made for them to report their status. Some designs provide for a status word to be associated with each channel in a unique location in memory. Whenever an interrupt is generated, the interrupt handler can determine which status word to inspect. This determination can be made because either the interrupt itself is channel specific or the address of the unit causing the interrupt is provided to the interrupt handler. Some systems have only a single status word that must be transferred to a safe location by the interrupt handler before the interrupt handler can allow interrupts due to I/O to occur again.

Channels that connect to control units may be *simplex*, *half-duplex*, or *full duplex*. A simplex channel has only one direction. It may transmit to a control unit or it may receive from a control unit. A half-duplex channel may receive or transmit, but not at the same time. A full duplex channel may receive and transmit at the same time.

## 8.  GENERAL-PURPOSE CHANNELS

A further generalization of interconnection uses a channel that is generalized so that any of a number of control unit types may be attached. All specialization associated with a device type is designed into the control unit that interfaces, through a standard interface, with a general-purpose channel. The channel may connect with a variety of different control units, and a system may have a multiplicity of channels. This design provides for enormous populations of devices to be attached to the system. An I/O address contains three components, the address of the channel, the address of the control unit, and the address of the device.

The degree of parallel I/O operation depends on the split of function between channel, control unit, and device. Consider an operation on disk that must first SEEK a disk address. If the SEEK address is kept in registers associated with the channel, only one SEEK operation per channel may occur. If the SEEK address is associated with a control unit, then one SEEK operation per control unit is possible, if the SEEK operation is associated with a device, so that the address is contained in register space within the device, then a SEEK per device may be undertaken in parallel.

Certain channel and control unit designs may enable more than one device to be actively transmitting or receiving data at the same time. A control unit associated with slow devices, like printers, readers, punches, or paper tapes, may time-divide multiplex among the slow devices, accepting characters from one device and transferring them to memory before another device delivers a character. Control units and channels that have this multiplex capability may also operate in *burst mode* for periods of time to service a faster device.

Another flexible feature of I/O design at this level allows control units to connect to more than one channel and devices to connect to more than one control unit. This provides alternate paths in case of malfunction and, if the operating system is so designed, the ability to address devices through the least busy paths. An ability to connect to up to four alternate units is common.

Each channel may be thought of as a specialized E-unit with sufficient logic to interconnect with control units, request memory fetches, and stores on behalf of the control units. From the processor point of view it is concerned with an I/O operation for only as long as it takes to deliver necessary function and control information to an indicated channel. It then experiences intermittent delay at the memory interface as channels and processor contend for memory. The job of fielding I/O completion or other kinds of interrupts, however, still rests with the processor.

Depending on the split of function between units on the pathway to a device, control information may be retained in a channel, passed on to a control unit, maintained in the control unit, or passed on to a device. The multiple channel/ multiple control unit design is typical of large systems which may have large numbers of channels. Though channels are generalized, there may be differing speeds and some specialization in the I/O design.

A goal of design is to maximize the potential flow of data to and from the memory in terms of characters or words per second. Some approximations of how high a data transfer rate a processor of a certain MIPS rate needs have been undertaken, but it is so subject to environment and use that a general transfer rate/MIPS rate ratio has not yet been discovered. Contemporary large systems have potential data transfer rates that are limited only by the speed at which the memory can support DMAC accesses.

The reader should be aware that some differences in the use of the word channel occur in the industry. IBM tends to use the word channel to designate a unit containing a good deal of staging and multiplexing logic that is really a small processor. Others tend to use the word merely to refer to the pathway, the bus, that connects processor and memories to control unit logic. Thus when IBM speaks of a processor with 12 channels, the capability of those channels may match the capability of a larger number of channels when the word is used to mean merely a data path.

Some real differences of design style are implied here. The most elaborate IBM I/O designs tend to provide for more complex control of data flow, subdividing the total data flow in response to spontaneous demands from devices and control units. Other designers tend to get large data rates from the I/O population by increasing the count of data paths so that the total I/O rate is achieved by parallel operation of channels whose maximum rate is below the maximum rate of an IBM style channel.

Clearly the efficiency of I/O depends on references to devices being spread across multiple channels and, perhaps, across multiple control units. Those responsible for configuring a system and those responsible for installing applications must consider the pattern of I/O references anticipated so that maximum I/O flow can be achieved. It is not desirable for large queues to form on a particular channel or control unit while other channels and control units are idle. The resource management algorithms of the operating system may help in allocating spaces to programs in order to maintain a balanced I/O flow.

A design that addresses the problems of bottlenecking and bad balance is called *floating channel*. Floating channel has the characteristic that a number of channels can reach the same control unit. There is no fixed relationship between a device and a particular path or set of paths to it. An I/O instruction names the device to be addressed. A switching unit selects, from a set of channels, a free channel that will lead to the device. All of the control units attached to a particular set of channels can be reached by any channel in the set. Consequently, any device can be reached from a set of channels. The channel used is the one first found free by channel selector switching circuits. The need for sophisticated software load balancing algorithms is considerably reduced, as is the care with which on-line data sets or files are assigned to particular devices.

## 9. INPUT/OUTPUT PROCESSOR CONCEPTS

Up to this point the overlap of I/O operations has been supported by a set of specialized E-units, sometimes called channels, that independently control the flow of data from I/O devices. The idea of read/write/compute overlap may be further supported by extending the power of the E-units until they effectively become separate processing units called *I/O processors* or *I/O managers*. Figure 32 shows a system with a processor and an I/O processor configured so that each has a path to memory. The I/O processor and the processor may have the same or different instruction sets and the split of work between them can vary widely. One design provides for the processor and I/O processor to share designated portions of memory. These shared addresses are used to form I/O request queues. Whenever the processor wishes to execute an I/O function it places a software defined packet in memory. The I/O processor inspects memory from time to time looking for I/O requests. When it finds a request the I/O processor submits it to the channels and control units to which it is attached. The processor has no notion whatever of I/O, can execute no I/O commands in

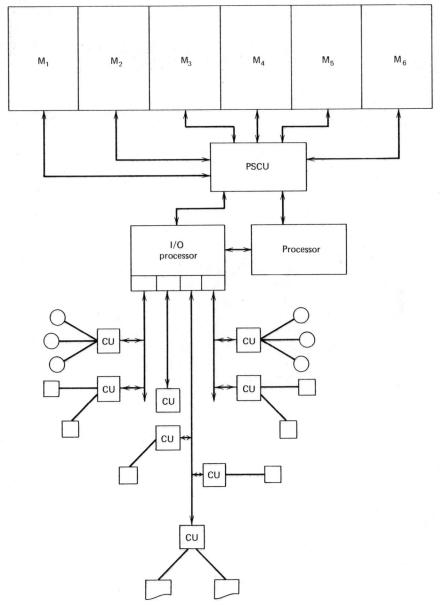

**Figure 32.** I/O processor.

either problem or control state, fields no I/O interrupts, and so on. The I/O processor attends to all these functions, providing the processor with a flow of data. When an I/O operation is complete, the I/O processor notifies the processor by some protocol that may involve an interrupt or merely the placing of a message in shared memory.

I/O processor function can be extended beyond what might be done by a processor in control state. The functions of record advance and buffer control, as well as record formatting, may be placed in the I/O processor. Notions of back-end data engines or front-end network controllers are basically extensions of the idea of an I/O processor where I/O functions performed on the auxiliary processor reach up to a level well beyond device control and DMAC transfers.

## 10. PROCESSOR SUPPORT OF INPUT/OUTPUT FUNCTIONS

As processor logic and memory become cheaper it is becoming common to implement control functions in devices by the use of small processors with associated control and buffer memories.

At various levels of a device hierarchy there may be buffer memories into which data are placed. In the hierarchy data may flow from buffer memory to buffer memory, under control of cooperating processors in the devices so that movement through the hierarchy does not cause data to flow through the memory of the main computational processor.

In such designs the local device processor may execute various algorithms for determining what data should be moved from the continuous motion part of the device (rotating drum or disk) into the static buffer.

In general we may expect variations of designs that put processor intelligence into basic I/O device design. In effect then, at some time in the future all I/O activity may be effectively processor-to-processor activity and there will be no dumb devices. This is clearly becoming true of terminals and disk units, and it is also a feature of some printers. The tendency is to move control functions closer and closer to the end of the data path to provide for maximum parallel operation and maximum I/O flow.

There is some concern about how very fast processors of the future can be fed with data quickly enough to sustain MIPS rates well beyond current processors. Design tends to maximize the offload of function and push I/O control to the edges of the data path. Thus small processors may be imbedded in the read/write heads of arms servicing disk units to assure maximum parallel data flow and controlled staging throughout many levels of hierarchy.

## QUESTIONS

1. What is the fundamental nature of programmed I/O?

2. What is the most basic overlap between processor and I/O?

3. How is overlapped I/O coordinated without an interrupt?

4. What is DMAC? What is the intention of DMAC?

5. What is the function of a control unit?

6. What is a half-duplex channel?

7. What is a general-purpose channel?

8. How does the split of function between elements along the I/O path affect performance?

9. What is the purpose of an I/O processor?

10. What is the use of processors imbedded in I/O devices?

**Chapter 21**

# Overview of Implementation

## 1. INTRODUCTORY OBSERVATIONS ON TECHNOLOGY

Those of us who work in the computer industry can scarcely pass a day without our attention being brought to the miracles of rapidly improving technology. The electronic components that will be available will be cheaper, faster, more densely packed (and consequently smaller), require less power, and so on. The promise of essentially limitless progress is constantly before us as technology marches on. There is almost no way, it seems, that a chapter on implementation issues that refers to available technology will not seem obsolete by the time a book is published.

It is interesting to take a short look at the progress technology has made since the early days of computers in order to lay the groundwork for a discussion of the methods that can be used to implement a machine. We will indeed perceive substantial change that will influence not only the design but also the architecture of computers. It stands to reason that if the cost of circuits and their packages becomes trivial and the circuit count of machines can dramatically increase with little increase in cost, then there will be a tendency to use the circuits to increase the speed of an architecture (enhance design) or to raise the level of an architecture in order to make it more programmable, more like accepted high-level computer languages, and so on. Different schools of architecture and design will argue about which approach is best and how to determine trade-offs, the proper rate of innovation, and so on, but computer architecture and design will naturally move forward in the presence of new technology.

### 1.1. Physical Change

The author's first machine was the UNIVAC (Universal Automatic Computer). The machine had a memory of 1000 12-character words, attached to a maximum of 10 tape drives. It could load a register in an average time of 445 microseconds,

**290**

do 12-character decimal addition in an average time of 525 microseconds, and decimal divide in an average time of 3890 microseconds. The notion of average time comes from the fact that the primary memory consisted of mercury delay lines that moved sound waves in mercury at a particular rate so that a memory reference was subject to some latency time. It could read/write/compute, using a buffering scheme for input and output of fixed size tape blocks. It was programmed primarily in machine code when machine code did not imply the intervening convenience of an assembly language (although what was perhaps the first work in macroassembly languages and COBOL-like languages was undertaken for the system). The UNIVAC filled a room, was water-cooled, weighed many tons, had something like 6000 tubes, and was physically entered for maintenance. The cost of the machine was millions of dollars and its operating console was larger in physical size than the processor and memory of a mid-sized computer today. (It was the glory of its time).

It could be embarrassingly outperformed (from the point of view of processor speed and memory size) today by a home computer costing well under $3000.

The vacuum tube, of course, is no longer the fundamental electronic building block of computers. We have passed to the age of the transistor and the integrated circuit, and we have passed from *medium scale integration* (*MSI*), with hundreds of transistors per chip, to *large scale integration* (*LSI*), with thousands of transistors per chip and are now confronted with *very large-scale integration* (*VLSI*) as the basis for conjecture that a chip holding a million transistors will some day be available.

Let us say a word about what a chip is. The basic electronic building block material is silicon. This *semiconductor* can be employed in various technologies called *bipolar* and *field-effect technologies*. The details of these technologies are beyond the intent of this book, the phrases refer to the fundamental properties of the electronics and physics that underlie computer technology.

Processed silicon is sliced and diced into very thin, characteristically square, units that have thousands (or more) transistors that can be used as memory cells or components of logical data flow. These units, well under fingernail size, are the chips. These chips have a set of I/O *pins* that allow the chips to be connected to each other logically and physically. A set of chips so interconnected are mounted on a unit called a *card* that varies in size but is commonly less than the size of the physical dimensions of the jacket of this book. These cards are themselves attached to boards that are collections of interconnected cards, and boards commonly attach to a bus that provides for a board-to-board connection that is necessary to create a computing system.

The physical appearance of a machine of the medium-sized IBM Series/1 class is shown as Figure 33. The figure shows a frame which has a mother bus running at its base. Plugged into this bus are the boards that comprise the system. A processor uses one to three boards to represent the processor logic and a number of boards for memory. In addition there are boards for standard and specialized I/O. The configuration of the system is limited by the capacity

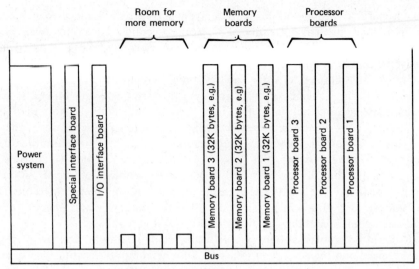

**Figure 33.** Small machine packaging.

of logic on each board and the number of boards than can be plugged into the bus.

All of the miniaturization of computers has been accomplished with silicon-based technologies. The industry is looking forward to a switch from semi-conductor-based technology to *superconductor*-based technology. One area of work is aimed at *Josephson junction technology*. This technology would drastically reduce power requirements for computers, operate at around the 30-picosecond level, and possibly provide an order of magnitude more MIPS in current generic architectures. The critical factors in the development of Josephson junction machines are (1) the miniaturization of the cooling components that must be used to obtain the switching and delay characteristics of the technology—the Josephson technology must operate at temperatures that are very low on the absolute temperature scale and (2) the shielding of electronic components from the sources of electrical activity that occur naturally in the environment. There is reason to believe that super-computers of the 80s may have some superconductor components.

## 1.2. Performance and Price/Performance

Associated with the increasing density of electronic circuits is a constant decrease in cost and increase in speed. We have progressed from milliseconds $(10^{-3})$, to microseconds $(10^{-6})$, to nanoseconds $(10^{-9})$, and are now entering the world of picoseconds $(10^{-12})$. The fastest of the commercially available computers as of this writing has a 12.5-nanosecond machine cycle. This implies basic logic cycles of speeds considerably higher. A large commercial machine, by no means intended to be a super computer, has a basic machine cycle of 24 nanoseconds.

From a performance point of view, we are being told that we are just about at the beginning. From a cost point of view we are being told very much the same thing. An interesting figure is the cost per instruction executed for the fastest available general-purpose machine at any time. This cost per instruction is calculated by dividing the MIPS rate into the cost of processor and memory. The figures shown are my (not IBM's) approximations for machines at the top of the recent IBM product line. MIPS rates are approximations and do not represent a statement of performance promulgated by the IBM Corporation.

| Model | Announce Date | MIPS Rate | Cost per Instruction ($) |
|---|---|---|---|
| IBM 370/168-3 | 3/75 | 2.8 | 1.60 |
| IBM 3033 | 3/77 | 4.5 | .74 |
| IBM 3081* | 11/80 | 6.0 | .40 |

The improvement in both performance and price/performance of these large machines is obvious. Since the late 1950s in fact there has been a trend toward constant improvement of price/performance, and most predictions are that this will continue with no decrease in rate.

One interesting detail is that the price effectiveness of the fastest general-purpose processor no longer necessarily indicates the best price/performance ratio during an interval. There is some argument over whether economies of scale exist for processors and memory. Those who believe that economies of scale are inherent in computing devices argue that the major components of computer cost in the future will lie in packaging, power, and cooling elements. The relative cost of circuitry to total systems cost will be low. Since there are economies of scale in power supplies, systems frames, and cooling mechanisms, there is maximum motivation for sharing the use of these elements among as many circuits as possible. Since increased circuit counts can be used to build faster machines by providing parallel execution, special-purpose execution units, buffers, pipelines, and so on, faster machines will be more price effective than slower machines.

There is a set of counter arguments that claim that economies derive from minimizing the number of unique components of a system and maximizing the replication of a minimum number of standard chips. This reduces design effort and expense and introduces an economy of scale in manufacture. A fast computer will necessarily be disproportionately more expensive, some claim, because a good deal of unique design and manufacture at low volume is necessary to produce a large system. The cost of the additional design and low-volume

---

* One processor. System is currently available only in two-processor configurations.

manufacture will offset the relatively low cost of increased circuit count. It is probably reasonable, for the near future, to assume flat pricing across a broad range of models of an architecture.

### 1.3. Storage

In addition to dramatic drops in the prices of logic and memory, there has been dramatic improvement in the storage capacity and price per megabyte stored for disk devices. There is also emerging technology to fill the *hierarchy gap*. The hierarchy gap occurs between main storage and disks and drums that are used as backing stores and I/O devices. The gap is that the speed difference between primary storage and these devices is much wider than one would wish for. The gap can be filled in two ways. One is to use memory-like devices that have multiple microsecond cycles. Another is to define a class of device that has a performance range in the higher microsecond range (perhaps 100s per access) but is very much cheaper than any main memory. Future hierarchies may have both levels added. It is absolutely critical that very fast machines can have larger memories and a memory hierarchy that can meet the enormous MIPS rates that may occur.

### 1.4. Summary Remarks

So wherever we look in the arena of computing hardware technology we find price decreases, increased capacity, and so on, with the promise of more and better to come. The impression that is formed by the constant rates of improvement is made more dramatic when one considers that prices and costs for componentry are commonly stated in terms of current costs without an adjustment for inflation. When we see that the cost of a stored bit drops dramatically from 1960 to 1985, we are looking at raw figures. If we impose on these figures the inflation rate during that interval, the results are indeed staggering. No other industry has maintained and sustained a record of increased power and reduced cost comparable to computer hardware since the Industrial Revolution.

### 2. COMPUTER BUILDING BLOCKS

All processors are combinations of a set of basic logical circuits of various types. Conceptually the basic logical elements of a computer are *AND gates*, *OR gates*, and variations of these. These building block elements are organized into combinations to form logical structures that determine the flow of data and control signals within the machine. An AND gate is a device that will emit a signal if it receives all of a set of designated signals as input. It derives its name from the requirement that signal 1 and signal 2 be present for the gate to emit a signal. An OR gate has the characteristic that it will emit a signal if any of a set of designated signals are input to it. Thus, if signal 1 or signal 2 is present, the

OR gate will emit an output signal. There are variations that involve the presence and absence of signals. These variations are called *NOR gates* and *NAND gates* and are widely used in computer design.

These basic logic elements are combined to form coding and decoding matrices, comparators, adders, complementors, and other higher-level computer structures. The advances of technology have not fundamentally changed the underlying logical building blocks of computer systems, although the packaging of these elements have been dramatically changed.

In the early days of computer design a logical designer uniquely designed each higher-level structure of a computer system using the basic gates as building blocks. Each logical element of a computer, the adder, the comparator, the I-box, and the E-box(es), had to be uniquely designed and translated into transistor populations that were organized on circuit boards; then the boards had to be wired together on a backboard.

The most important effect of miniaturization and increased speeds on computer implementation is that it is now possible to build very capable computers whose fundamental building blocks are themselves computers. The concepts of *microcode, firmware*, and *microprocessors* are important in this area.

Complete logical design at the gate and circuit level is now absolutely necessary only in the area of building the smallest and largest computers. The smallest computers require this kind of design because there is no other underlying building block to use. Large computers require this design because of the desire to create organizations of enormous complexity. Computers in a broad capacity range can be realized by the use of smaller, less complex underlying computers that run programs that define the architecture of a *target machine* with a more capable instruction set, richer addressing conventions, and so on. These support computers, of course, may sometimes be especially designed at the logic design level.

When the implementor of a computer uses standardized small processors (microprocessors) to reflect the architecture of a *target machine*, his decisions involve how many small processors to use, how to split functions across the processors being used, and how to organize memory. The implementor of a machine is basically making program organization and configuration decisions about the computer system he will use and the program he will write to create an application that is another machine. He decides what basic chips to use at the level of choosing between standardized small processors with different memory sizes, and whether to use just one type or a number of processors available in a *chip family* that may consist of a basic processor, an I/O control processor, a specialized processor for floating point, and so on.

An implementor of computers may not use just standard small processors but may design his own processor to support the architecture of a target machine. In this case he must specify the architecture and the design of the underlying machine. The instruction set for this underlying processor is frequently called *microcode*. The implementor trades cost against speed in order to determine how much of the target machine's architecture and function will be programmed in the underlying machine, how much specialized hardware should be used, and, in general, how efficiently the target machine can be represented in the underlying machine. Slower, less expensive versions of the target machine will have less hardware capacity in the underlying machine and rely predominantly on programs running in the memory of the underlying machine to represent the target machine. Faster, more expensive versions of the target machine will use more hardware support. For example, the width of the data paths of the underlying machine may be made the same as the data paths suggested by the architecture of the target machine, MULTIPLY and DIVIDE may use some specialized hardware rather than programmed algorithm in the underlying machine, floating point hardware may be made available, and so on.

So, although the basic building blocks of computers continue to be the fundamental logical functions, the packaging of these units in such density has provided a new set of choices for the designer of machines. He now has building blocks available that are themselves small computers and can make use of these computers in the implementation of larger computers. In effect he can choose how much of a machine will be hardwired and how much implemented in algorithm represented by programming that emulates the architecture of a target machine. In addition to the design decisions of the split of I-time and E-time functions, the width of the data flow, and the split of functions across components, there are now additional decisions at a new level of hardware/software interface that one may call the hardware/firmware interface.

The hardware/software interface is essentially the issue of the split of function between the target machine and its software. The hardware/firmware interface is the issue of the split of function between the hardware of the underlying machine and its own software in the emulation of the target machine architecture.

An additional level of consideration exists in the fact that the underlying machine may itself have an underlying machine, so that there are three levels of architecture, design, and implementation.

## 3. ARCHITECTURE, DESIGN, AND IMPLEMENTATION

The development of a machine of a certain performance and price/performance is influenced by the architecture. We have earlier discussed some of the influences that architecture and design have on each other. In order to achieve certain speeds for an architecture within a technology period that determines the basic switching speeds, propagation delays, and so on, it may be necessary to use rather elaborate designs involving buffering, pipelining, and parallel execution.

In Chapter 12 we saw attempts to derive measurements for architectural power that are design and implementation independent. One important measure of an architecture must naturally be how complex the design must become to achieve certain levels of performance.

Clearly the complexity of a design required to run an architecture at a certain speed is very much influenced by the underlying technology used to implement the machine. Only if performance that exceeds the raw capability of the technology is required must designs be undertaken to increase the instruction retirement flow by increasing the number of parallel executing components or by elaborating the design of a particular critical function.

A clear alternative to design complexity is the choice of a technology so fast and dense that simpler designs may achieve faster speeds. This choice of a faster technology may involve some risks as the use of very aggressive technologies may introduce more complex problems in component test, manufacture, cooling, and powering. The payoff for using more aggressive technology, however, is that the cost of design may be reduced by reducing the complexity of the design. Fewer unique components may be required, buffering requirements may be reduced, and so on. The issue of future economies of scale relates to the balance between fast technology and elaborate design.

In the same spirit, faster technology may allow the more aggressive use of circuits to support higher-level architectures rather than to achieve concurrency and overlap.

A vendor of computing systems must now determine how aggressive and innovative he wishes the architectures to be, how aggressive he will be in the choice of underlying technology, and how he will trade off design sophistication in support of speed or architectural level. It is going to be an interesting decade.

## 4. MICROPROCESSORS AND MICROCOMPUTERS

It is very difficult to provide a precise definition for a microprocessor, but a reasonable rule for recognition is that all of the logic for instruction preparation and execution exists on one chip. As this is written the number of circuits on a chip to support the architecture of a microprocessor ranges from around 4K to 64K.

Despite their small size, contemporary microprocessors are by no means architecturally or design naive. Much work in achieving higher-level architectures is associated with microprocessors. For example, the INTEL 432 has undertaken an architecture extended toward the features of a subset of the ADA language.

Concepts of lookahead, virtual memory, and parallel execution are beginning to be associated with microprocessor development. There are a number of 16-bit microprocessors now available, and we have every reason to believe that even wider data paths will be available at this level. So we are dealing with a basic instrument whose fundamental characteristics are fast approaching or exceeding those we associate with intermediate or large computers currently installed in typical data processing environments.

One use of a microprocessor is to build a microcomputer. A microcomputer is generally a multichip system that provides all computer functions including interfaces to I/O. A microcomputer may contain one or more microprocessor chips and a set of memory and specialized I/O interface chips on one card or on multiple cards. The cards are purchasable without further packaging, as are the component chips, and microcomputers are also commonly available in packaged versions with a simple console.

It is quite common to find microcomputers packaged and marketed as small business computers, personal computers, and so on, where they are offered with a display tube and keyboard, some diskette storage, and a printing device of some kind. Trade names like IBM, APPLE, COMMODORE, and TRS-80 are currently widely known packages of computer power based on microcomputers.

A microcomputer may look like Figure 34. Here on a single board, sometimes marketed as a *single board computer* (*SBC*), for example, the INTEL 8030, are the chips necessary to provide a complete computer system. We have

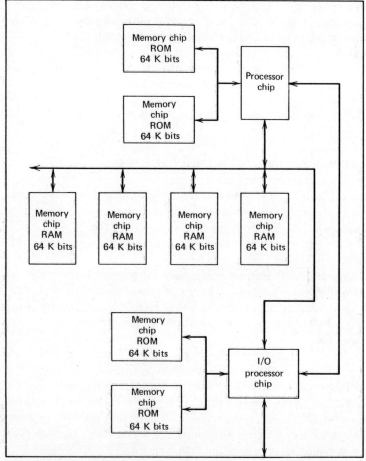

**Figure 34.** Simplified conceptual single board computer.

a microprocessor chip that contains registers, control and arithmetic circuitry, and a *random access memory* (*RAM*) chip that contains memory locations available to any program running on the microprocessor. The RAM chip is what we would think of as "normal memory," it can be read and written and executed from.

In addition, the figure shows a *read only memory* (*ROM*). A ROM memory is a memory that cannot be changed by a program running on the microprocessor. The ROM may be used as a specialized part of accessible general memory that cannot be changed or it may be used as a special control store to contain, for example, coding that enhances the architecture of the microprocessor. A floating point subroutine may be contained in the ROM so that whenever a program issues a floating point function it is executed in the ROM memory. A faster floating point function might be optionally provided by a floating point chip that contains hardware to execute floating point functions.

There are various versions of ROM type memories. An *erasable programmable read only memory* (*EPROM*) is a ROM that can be changed by removing it from the computer and passing it through a process that will clear its contents and place new contents in the locations. A device that does this is commonly called an *EPROM burner*.

It is now becoming common to use a *writable control store* where nonwritable stores have been used in the past. The contents of such a memory can be changed by specialized procedures, for example, the loading of the contents of a diskette as a result of a privileged operation by an authorized operator of the system.

The last element of Figure 34 is an I/O chip. This chip shares RAM memory with the microprocessor and connects the microcomputer to peripheral devices.

The microcomputer we have shown is rather basic. Much more complex systems structures can be built from the building blocks. These include systems with more than one I/O processor, with more than one computational microprocessor, and with an ability to define commonly addressed memory shared between the processors.

## 5.  BASICS OF IMPLEMENTING ONE ARCHITECTURE WITH ANOTHER

We are going to show a very basic one-level implementation of a target architecture with a collection of microprocessors. Consider Figure 35. This shows a system that has three microprocessors, a set of RAM chips, and a set of ROM chips.

Microprocessor A is dedicated to the execution of the I-functions of a target architecture Microprocessor A has access to ROM 1 and to some of the RAM chips. The shared RAM chips represent the memory of the target architecture and contain all data and programs that exist on the apparent machine. ROM 1 contains coding in the language of microprocessor A that represents the archi-

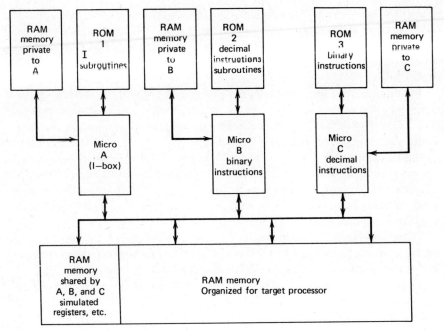

**Figure 35.** Underlying machine structure.

tecture of the target machine I-functions of instruction fetch, decode, form addresses, and so on.

ROM 1 may contain a set of subroutines that have titles like ADVANCE INSTRUCTION COUNTER, ADD BASE REGISTER, and ADD INDEX REGISTER. Just what microprocessor subroutines exist in the ROM 1 memory depends on the details of how the work is split among microprocessors A, B, and C. Earlier in the book we mentioned that there are some options about the definition of what the I-time and E-time functions are. Requesting an operand, for example, may be an I-time or E-time function.

The processing of the I-time functions of the target architecture consists of a sequence of subroutine calls that:

1. Advance a RAM 1 memory location representing the instruction counter of the target machine.
2. Bring a target machine instruction in RAM to microprocessor A.
3. Analyze the preparation requirements for the target instruction.
4. Form target instruction operand addresses.

All of the I-time phases are performed by executing microprocessor subroutines that perform on the target machine instruction exactly as if the logic of I-time were hardwired into the system. The ROM memory particular to microprocessor A contains the instructions it executes; the RAM memory particular to A contains normal addressable working space needed during execution.

When microprocessor A is finished with the I-time functions, it passes an instruction on to microprocessor B or C. It makes a determination of which E-time microprocessor to pass an instruction to during the DECODE subroutine. We use B to process all binary instructions and C to process all decimal instructions.

A message containing functional description and operand address is passed to B or C by use of a designated area in a shared RAM. Part of one RAM memory contains bytes that are not accessible to a program running on the target architecture. These bytes are used for communication between the three microprocessors. Another use of this special memory area may be to simulate the registers of the target architecture so that, for example, target architecture index register 1 is in memory locations 0056–0059.

When an E-time processor receives an instruction in shared RAM it undertakes to execute it by executing code in its dedicated ROM that represents algorithms appropriate for the execution of the target architecture instruction. It collects operands and uses its own arithmetic capabilities to perform functions that will end in an appropriate result, which is then placed, for example, in the RAM memory locations representing the registers of the target machine.

The programs in the ROM memories associated with the E-time microprocessors may be organized in various ways. Most straightforward, perhaps, is to have a program representing each of the instructions that the particular E-time microprocessor can perform. This set of programs will make calls to subroutines that represent functions that the instructions may have in common, such as bring an operand to a register.


## 6.   CONSIDERATIONS IN IMPLEMENTING ONE ARCHITECTURE WITH ANOTHER

The underlying microprocessor support of a target architecture in the previous section used three microprocessors to achieve the architecture. The architecture might have been supported on one. Motives for using more elaborate underlying machine structures may be based in program organization considerations and in performance considerations. Given constraints in addressing for a particular microprocessor, it might be more convenient to organize I and E-time operations in a way that makes manipulation by three processors convenient. Given three processors, overlap relations can be defined between them so that significant lookahead or even parallelism can be achieved.

Using contemporary off-the-shelf microprocessors to support a target architecture, however, will still lead to target architecture performance of reasonable modesty. A one-MIP performance rate for a target architecture of the nature of an IBM/370 is probably not achievable. Using underlying machines to simulate faster versions of a target architecture is usually approached in other ways.

One approach is the use of *bit-slice technology*. Bit-slice is a methodology whereby the data paths of an underlying machine can be matched to the data path of the target architecture and a high degree of parallelism achieved.

In bit slice organization each chip is a small processor with an instruction set, a control memory, and a very narrow data path, of perhaps 2 bits. A 32-bit architecture is achieved by placing 16 of the 2-bit slices in tandem and coding each processor so that it is aware of its interconnection with another slice. Thus a 32-bit add can be performed on 16 2-bit processors, each processor containing the coding for a 2-bit add and coding to accept carries from a processor at its side.

The presence of the family of bit-slice processors provides the ability to effectively map the underlying machine to the data flow of the target machine and to achieve speed by wide data paths as well as by the degree of parallelism between the processors. Bit-slice processors of various widths are available as off-the-shelf items, for example, the American Microdevices 2900 family. Many medium-sized computers that are implemented using underlying processors are based on a bit-slice implementation.

Many larger models of target architectures are built from underlying processors that are specially designed for the purpose of emulating the target architecture. Each model of the target architecture may be supported by a different underlying processor designed to provide the price/performance goals of a particular model of the target architecture. In the family of generic IBM/370 architecture machines, for example, each IBM-produced model is supported by a different implementation using different underlying processors and different balances of hardwiring and microcode. Faster models are supported by more hardware and underlying processors with richer instruction sets and good mappings of the data paths; slower models are supported by less elaborate, and consequently less expensive, underlying processors. Generally such underlying processors are not considered to be microprocessors, and the phrase "micro-coded machines" is used to distinguish implementations with unique underlying processors.

The design of the underlying processor involves a number of price/performance decisions that include the degree to which the underlying machine should be general-purpose or specialized. An underlying processor may be designed to support more than one architecture or performance level. In this case the data flow would be as generalized as possible, and the register set, addressing, and so on would be organized to allow reasonable emulation of a variety of architectures. The program in the underlying architecture would bear most of the burden of simulating the selected target machine register population, addressing, and instruction set.

An underlying processor aimed at a specific model of a specific target architecture would have many of the features of that target architecture built into it. For example, the register model of the target architecture might be directly represented in the architecture of the underlying machine. This would make the emulation of the target architecture more direct with less work done by programs running on the underlying architecture.

One characteristic of the design that occurs in microcoded or microprocessor supported architectures is that the concept of the basic machine cycle or the basic logic cycle becomes a little vague. Since the basic cycles are really only relevant to the underlying machines, the basic cycles of the target machines may not be as precise and orderly as they are on a hardwired target architecture. Timings may reflect variations that result from the dynamics of underlying machine operation, and it may be difficult to state them exactly in terms of target machine cycles.

## 7.  MICROCODE TYPES

One of the design decisions for an underlying processor is the nature of its instruction set. It is important to be familiar with the concepts of *horizontal microcode* and *vertical microcode*.

Vertical microcode is a characterization of instruction sets of underlying processors when these instruction sets look very much like instructions of the classical type. The vertical instruction set is a set of operation codes used in instruction words that is formed very similarly to the way in which the instruction sets we have discussed throughout the book are formed. Each instruction has a single operation code and some set of addresses. The emulation of the target architecture is accomplished by the execution of programs in the memory of the underlying machine in a way that is rather familiar to us. As we have seen, the efficiency of the emulation depends on the degree to which the underlying machine is supported by hardware and on the number of instructions required to represent an instruction in the target machine.

In systems where vertical microcode is used, the concept of microcode is almost more a legal than a technical concept. Microcode, a lawyer will tell us, is code that is protected by a vendor and not made available in the documentation and representation of the system. The intent of the vendor in so restricting this information is to assure that the machines that he sells are not modified in ways that make them impossible to maintain.

The phrase "horizontal microcode" refers to an alternative organization of an underlying instruction set so that microcode looks less like a typical small machine instruction set. In horizontal microcode, the instruction (microcode) word is a rather long word used to contain a number of operand directives and addresses. The intent is to be able to accomplish as much of the target machine logic as possible in parallel. The effect of such a long word is to reduce the number of microcode instructions necessary to interpret an architecture thereby reducing the number of sequential operations involved in architectural simulation.

In an extreme form of horizontal microcode, each gate of the architecture may be represented by a bit position in the horizontal microcode word that determines whether a signal will or will not emit from that gate during the execution of the horizontal microcode instruction.

One advantage of horizontal microcode is that it can be used to design data flow and control almost with the efficiency of hardwired logic because of the extremely high parallelism that can be represented in the microcode word. A disadvantage of horizontal microcode is that rather simple functions, which cannot make use of the potential, require disproportionately large amounts of effectively unused space in a microcode word.

In order to achieve some balance between horizontal and vertical approaches, many machines have a level of horizontal microcode and a level of vertical microcode used to support the target architecture.

The instruction set of the underlying processor is vertical, but beneath the vertical instruction set there is a horizontal instruction set that supports it. Thus three levels of architecture are involved in the support of a target machine. There are target machine architecture, the architecture of the vertical underlying machine that emulates the target machine, and a horizontal instruction set that performs the functions of the vertical set of the underlying machine. When two levels of microcode are used in this way, the base level is often referred to as *nanocode*.

## 8.  EXTENDING AN ARCHITECTURE

Microcoding underlying processors, microprocessors, and similar approaches may be used not only to simulate a target architecture but to extend the target architecture. Much of the work at the hardware/software interface and in the general area of higher-level machine architectures involves the use of microcode, microprocessor-based emulation of higher-level architectures.

Currently many vendors are analyzing the functions of operating systems to determine what software functions currently delivered in the operating system or other software may be placed into microcode or implemented on a processor. The placement of frequently used operating systems functions into microcode in order to speed their performance is sometimes called *microcode assist*. Microcode assist for operating systems functions may be used extensively even in systems where the fundamental architecture is essentially hardwired.

In the area of support of higher-level languages we seem to be entering a new phase in which the relationship between interpreters and compilers is being reinvestigated in the light of current microprocessor and microcode technology.

Since this is an issue we introduced in Chapter 1, we have obviously come full circle and this introduction to computer architecture and organization is now complete.

## QUESTIONS

1.  What is a target machine?

2.  How has progress in technology changed how machines may be implemented?

3. What is the nature of a design/technology trade-off?

4. What is a microprocessor?

5. What is a single board computer?

6. What is bit slice?

7. What considerations apply in the design of an underlying processor?

8. What is horizontal microcode?

9. What is nanocode?

10. How does microcode or microprocessor implementation relate to concepts of the level of an architecture?

## BIBLIOGRAPHY

This bibliography provides source material for Chapters 13–21.

*Communications of the Association for Computing Machinery* is shown as *CACM*; *Computer Architecture News* is shown as *CAN*; *Proceedings of ACM Symposia on Computer Architecture* are shown as *ACM-SIGARCH*.

Agerwala, T. K., K. M. Chandy, and D. E. Lang. A Modeling Approach and Design Tool for Pipelined Central Processors, *ACM-SIGARCH, Proceedings 6th Annual Symposium on Computer Architecture*, April 1979, p. 122.

Bell, R. K., et al. The Big Three—Today's 16-Bit Microprocessor, *SIGMICRO Newsletter*, Vol. 11, Nos. 3 and 4, September–December 1980, p. 126.

Bremer, J. W. Hardware Technology in the Year 2001, *IEEE Computer*, Vol. 9, No. 12, December 1976, p. 31.

Chen, T. C. Overlap and Pipeline Processing, *Introduction to Computer Architecture*, H. S. Stone, Ed., Systems Research Associates, Chicago, IL, 1975.

Davidson, E. S., A. T. Thomas, L. E. Shar, and J. K. Patel. Effective Control for Pipelined Computers, *Proceedings IEEE COMPCON*, Spring 1975, p. 181.

Feth, G. C. Progress in Memory and Storage Technology, IBM Research RC5881, Yorktown Heights, NY, February 1976.

Gabet, J. VSLI: The Impact Grows, *DATAMATION*, Vol. 25, No. 7, June 1979, p. 108.

IBM Corporation, IBM 3081 Functional Characteristics, GA22-7076, November 1980.

IBM Corporation, IBM 3033 Processor Complex Functional Characteristics, GA22-7060, April 1977.

IBM Corporation, An Introduction to Microprogramming, GF20-0385, December 1971.

IBM Corporation, A Guide to the IBM 3033 Processor Complex, Attached Processor Complex, and Multiprocessor Complex of System/370, GC20-1859, August 1980.

Kleinsteiber, J. R. IBM4341 Hardware/Microcoder Trade-Off Decisions, *SIGMICRO Newsletter*, Vol. 11, Nos. 3 and 4, September–December 1980, p. 190.

Patel, J. H. Pipelines with Internal Buffers, *ACM-SIGARCH, Proceedings 5th Annual Symposium on Computer Architecture*, April 1979, p. 249.

Patterson, D. A., and C. H. Sequin. Design Considerations for Single-Chip Computers of the Future, *IEEE Transactions on Computers*, Vol. C-29, No. 2, February 1980, p. 108.

Ramammoorthy, C. V., and H. F. Li. Pipeline Architecture, *ACM Computing Surveys*, Vol. 9, No. 1, March 1977, p. 29.

Reddi, S. S., and E. A. Fuestel, A Conceptual Framework for Computer Architecture, *ACM Computing Surveys*, Vol. 8, No. 2, June 1976, p. 277.

Riseman, E. M., and C. C. Foster. The Inhibition of Parallelism by Conditional Jumps, *IEEE Transactions on Computers*, December 1972, p. 1405.

Roooohu, W. G., and E. S. Lee. Performance Enhancement of SISD Processors, *ACM-SIGARCH, Proceedings 6th Annual Symposium on Computer Architecture*, April 1979, p. 216.

Sherwood, H. F. Technology of the 1980s and Organizational Issues, *BITS*, H. F. Sherwood & Associates, SA, Hamburg, W. Ger., Fourth Quarter, 1980, p. 14.

Smith, A. J. Directions for Memory Hierarchies and Their Components: Research and Development, National Technical Information Service, LBL-8276, UC-32, October 1978.

Sperry Rand Corporation (at time of publication, Remington Rand Inc., Echert-Mauchly Division), Programming for the UNIVAC Fac-Tronic System, EL192B, January 1953.

Towsley, D. F. The Effects of CPU:I/O Overlap on Computer Systems Configurations, *ACM-SIGARCH, Proceedings 5th Annual Symposium on Computer Architecture*, April 1978, p. 238.

Trivedi, K. S., and T. M. Signon. A Performance Comparison of Optimally Designed Computer Systems with and without Virtual Memory, *ACM-SIGARCH, Proceedings 6th Annual Symposium on Computer Architecture*, April 1979, p. 117.

# INDEX

Access methods, *see* Input/Output
Accumulator architecture, 37, 39–40, 157, 218
Activation record, 102, 130
Address formation, 190, 191, 193, 198
Addressing architecture:
  conventions, 5, 6, 22, 33, 37, 57, 61, 117
  indirect, 69, 98
  issues, 57–58
  linear virtual memory, 119, 121
  program, relative, 22
  three address forms, 61, 80, 157
  two address forms, 61, 80, 157
  virtual addressing, 70, 119–131, 122, 139,
    150, 151, 152
  *see also* Base registers; Index registers;
    Protection; Segment registers; Segments
AND, function, *see* Logical instructions
Arithmetic instructions, 75
Arithmetic registers, 6, 183, 185, 202–203,
  207
Architecture:
  criteria for efficiency, 149–154, 155–156,
    179
  definition, 3–4
  high level languages and, 4, 158, 164, 165,
    166–7, 304
  ideal, 158–160
  operating systems and, 165–166
  organization and, 10–12, 161–162, 173–174,
    177–178, 296
  power of, 147–168
  target, 295, 304
  *see also* Accumulator architecture; Army-
    Navy study; Operand stacks; Register
    file; Store-storage architecture
Army-Navy study, 148–154
Arrays, 34, 63–66
Assembly Language, 3, 8
ASCII, 30. *See also* Binary coded decimal
Assignment Statement, 19–23
Associative Memory, 127

Base registers, 37, 58, 62, 66–68, 106,
  183–185, 200, 202, 207
Basic logic cycle, *see* Logic cycle
Basic machine cycle, *see* Machine cycle
Binary coded decimal, 5, 28. *See also* ASCII;
  EBDIC
Binary representation, 5, 27–29
Bits, concept, 5, 27
Bit slice, 302
Board, 180, 291
Boolean operations, *see* Logical instructions
Branch, 7, 67, 77, 86–93, 209, 227–228,
  256–259. *See also* Condition codes
Branch/Link, 96–98
Branch prediction, 258, 259
Buffer management, 141. *See also* Instruction,
  buffer; Short made loop
Buffers, 210–211, 220, 227, 233, 251, 255.
  *See also* Instruction, buffer
Burroughs B1700, 168
Burroughs B5000, 45
Byte, concept, 5, 33

Cache:
  benefits, 260
  block address register, 264–265
  contents management, 265
  contents replacement, 265
  effectiveness, 261–262
  hit ratio, 260
  IBM 3033, 262, 264–266
  Input/Output and, 267
  memory mapping, 262, 264
  partitioning, 264, 266–267
  sizes, 266
  store through, 266
CALL, 77, 96, 99, 101
Capabilities, 129–130
Card, 180, 291
Channel, *see* Input/Output
Channel address word, *see* Input/Output

Channel command word, *see* Input/Output
Chaotic execution, 227–228
Chip, 180, 291, 295, 298, 299
Circuit density, 180
Circuit sharing, 174
Command chaining, 139
Compare, *see* Branch
Compiling, 13, 17–25, 216
  architectural impact, 161, 162
  assignment statement, 19–23
  code optimization, 23
  operand stacks, 18, 43–52
  operator stacks, 19–21
  postfix notation, 49
  prefix notation, 49
  quadruples, 20–23
  symbol tables, 18–20, 24
Condition codes, 80, 89, 112
Control:
  circuits, 183–185
  information, 55–56
  language, 143–144
  memory, 268–269
  registers, 6–7, 38, 54, 68, 117
  states, 9, 78, 116–118
  unit, *see* Input/Output
  *see also* Base registers; Decode output
    register; Index registers; Instruction
    counter; Instruction register; Memory,
    registers; Page registers
CRAY-1, 40, 52, 81, 84, 115–116
CYBER, 41, 80, 84

DASD, *see* Disk
Data:
  chaining, 139
  declaration, 27
  movement instructions, 74
  paths, 174, 182, 187–188, 247, 296
Decode, 7, 190–194, 198–200, 206, 211, 226,
  233–234, 256
Decode output register (DOR), 207, 209–211,
  216, 217, 219, 222, 251
Demand paging, 123–124
Descriptors, 35
Design, *see* Organization
Direct memory access (DMAC), 281
Disk:
  architecture, 134
  cylinder, 136, 252
  devices, 133
  rotational delay, 135
  seek time, 135, 284

  tracks, 134, 252
Dispatching, 42
Distributed processors, 233
Double word, concept, 33

EBCDIC, 30. *See also* Binary coded decimal
E-Box, *see* Execution units
Economies of scale, 293
Editing instructions, 77
Erasable read only memory (EPROM), 299
E-time, 189, 197, 207, 300
Exceptions, 107
Execution units (E-units), 175, 207, 209–211,
  215, 222, 227, 233, 236, 285
  DORs within, 217, 251
  functional grouping of, 217, 219
  heterogeneous sets of, 216, 217, 227
  homogeneous sets of, 215, 217
  multilevel units, 229
  populations of, 216, 217
  relations with registers, 219
  single level units, 228–229

File allocation, 140–141, 144
Firmware, 15, 295
First level interrupt handler, 108, 109
Flip-Flop, 180
Floating point, 5, 30–31, 150, 175, 177
Floating point operations per second, 231
Functional instructions, 155–156, 209

General purpose registers, 38

Halfword, concept, 33
Hardware/Firmware Interface, 296
Hardware/Software Interface, 13, 163, 296
Hexadecimal, 30
High level architecture, *see* Architecture
High level languages, *see* Architecture

IBM 3033, 231, 264, 266, 293
IBM 3081, 231, 293
IBM 4341, 260
IBM 8100, 43, 116
IBM S/38, 15, 128
IBM S/370, 11, 38, 40, 62, 84, 89, 90,
  110–112, 117, 128, 137, 142, 242
IBM S/370-168III, 293
IBM Series/1, 84, 103, 104, 116, 291
I-box, 206, 207, 209–210, 211, 227, 233–234,
  236, 253
ILIAC IV, 231

Incarnation record, *see* Activation record
Index registers, 6–7, 37, 52–53, 62–66, 77,
    184–185, 202, 207, 275
Indirect addressing, *see* Addressing architecture
Input/Output:
   access methods, 140
   address register, 278
   architecture, 10, 78, 107, 108, 117, 123,
     132–145, 150, 153, 271–272, 280
   blocking, 276
   buffering, 281
   channel, 137, 280
     duplex, 284
     floating, 286
     general purpose, 284–285
     half-duplex, 284
     simplex, 284
   channel address word, 137–141, 142–143,
     280–281
   channel command word, 137–141, 142–143,
     280–281, 284
   control unit, 137, 285
   count register, 279
   design, concepts of, 271–272
   devices, 132–133, 294. *See also* Disk
   directives, 136
   errors, 274, 276
   overlap, 233, 276–280, 281, 284
   programmed, 272–275
   processors, 286–288
   speeds, 238–239, 276
   START I/O, 137, 142–143, 280
   supervisor, 141–143
   TEST I/O, 276
Instructions, 6, 8, 20–23, 38, 57, 74–84, 150,
    152, 163, 167, 184
   buffer, 252–256. *See also* short loop mode
   counter (IC), 7, 54–55, 67, 184, 185, 197,
    198, 200, 202, 207, 236
   format, 79
   immediate, 77
   logical, 75–76
   overlap, 197. *See also* Parallel operations
   pre-processing function, *see* I-box
   privileged, *see* Control, states
   rearrangement, 227–228
   register (IR), 7, 200, 211, 218, 236, 242,
    251
   sequencing, 214, 216, 220–223, 226
   stages, 178, 182, 186, 189–190, 194,
    229–230
   times, 190
   types, 74–76, 155–156

Integration:
   large scale (LSI), 291
   medium scale (MSI), 291
   very large scale (VSLI), 291
INTEL 432, 129, 297
INTEL 8080, 81, 84, 99
INTEL 8086, 42, 68
Interpreters, 15, 22
Interrupt:
   classes, 110–111
   inhibition, 109, 110
   memory locations, 109, 110, 113
   priority, 116
   response, 107–108, 114–115
   status word, 108
Interrupts, 9, 105, 107–116, 150, 153, 284.
    *See also* Exceptions; First Level
    interrupt handler; Traps
I-time, 189, 197, 207, 300
I-unit, *see* I-box

Jump, *see* Branch

Linkage stack, 99, 100, 103
Loader, 25
Logic cycles, 180, 182
Lookahead, 197, 206, 210, 214, 221–222, 239,
    241, 243–244, 258

Machine cycles, 181–182, 186–187, 189–190,
    195, 205, 229
Memory:
   access time, 244
   banking, 67, 240, 245, 248
   blocks, 122
   boundaries, 242
   countention, 204–206, 210, 223, 226,
    239–240, 243, 278, 281
   cycle time, 244
   granularity, of address, 242
   hierachy, concept, 251, 267–268
   interconnect, 238, 245
   interface, 203, 245. *See also* Memory
    registers
   interleave, 240–242, 243, 245, 248, 260
   location, concept, 5, 31
   lookahead, 241, 242, 260
   mesh interconnect, 249
   modification instructions, 77, 78
   multiplexing, 249, 285
   multiport, 248, 249
   partitioning limits, 243–244

primary storage control unit (PSCU),
    245–247, 281, 284
reference times, 187, 198, 238, 241, 244,
    251, 256
registers:
    memory address register (MAR), 204,
        207, 211, 245
    memory data register (MDR), 198,
        202–204, 207, 211, 245
    memory fetch register (MFR), 204, 207,
        209, 211, 245
    memory interface register (MIR), 183,
        185, 198, 200, 202–204, 245
    memory storage register (MSR), 204, 207,
        211, 245
    operand request register (NDR), 245
    operand return register, (NDT), 245
    operator request register (OPR), 245
    operator return register, (OPT), 245
    scratchpad, 259
    sizes, 238–239, 269
    speeds, 205, 269
    star interconnect, 249
    store operations, 202
Microcode, 14, 162, 269, 295, 302
    assist, 304
    horizontal, 303
    vertical, 303
Microcomputer, 297
Microprocessors, 12, 297, 301
Microprogramming, 15
MIMD machines, 233
MULTICS, 71
Multiple I-streams, 233, 235
Multiple register operations, 39
Multiple register sets, 42–43
Multiprocessing, 150, 233
Multiprogramming, 9, 42, 236


NAND function, *see* Instructions, logical
Nanocode, 304
Non-functional instructions, 155–156
NOR function, *see* Instructions, logical
NOVA, 82–83, 92–93


Objects, 120, 129, 167
One's complement, 29
Operand forwarding, 218, 259
Operand registers, 6, 38
Operand stack architecture, 43–47
Operand stacks, 18, 43–47, 99
Operating systems, 10, 15, 25, 42, 72, 117,
        121, 123, 126, 130, 140–143, 165–166

Operation codes, *see* Instructions
Operator stacks, 19–21
Optimizing code, 23
OR function, *see* Instructions, logical
Organization:
    architecture and, 10–12, 161–162, 177–178,
        296
    concepts and definition, 173–179


Page:
    faults, 125–126
    management, 126
    registers, 66–68
    relative address, 67
    selection, 67
    tables, 122–123, 125
    virtual memory, 120, 122
Parallel machines, 214, 231
Parallel operations, 175, 178, 189–190, 191,
        193–194, 197, 200, 206, 224–225, 231,
        239. *See also* Instructions, stages;
        Lookahead
Parameter passing, 101, 103
Pipelined machines, 214
Pipelines, 229–230, 231
Postfix notation, 49
Prefix notation, 49
Price/Performance, 173, 176–177, 187, 231,
        269, 292
Procedures, 94
Program control block, 42
Programmable read only memory (PROM),
    269, 299
Program relative addressing, 22
Program status word, 89, 91, 111–112, 117
Propagation delay, 180
Protection, 70, 105, 127, 150
Protection keys, 71–72, 139
Protection rights, 71–73


Quadruples, 20–23
Queueing, 175


Random access memory (RAM), 299–301
Read only memory (ROM), 269, 299–301
Reentrance, 102
Reference, locality of, 124
Register architecture:
    in addressing, 37
    models, concept, 6, 22
    multiple sets, 42–43
    operand, 6, 38
    operations on, 39, 41
    usage status of, 221–222

Register file architecture, 39, 157, 218
Relative magnitude, 86–87
RETURN, 77, 96, 100
Return point, 97

Segment registers, 62, 66–68, 106
Segments, 68, 120–121, 127, 129–130
Self-describing data, 36, 167
Shifting instructions, 77
Short loop mode, 257
Skips, 92–94. *See also* Branch
SIMD machines, 231
Single board computer, 298
Single level store, 133
Software, 12
Stack pointer, 99
Stacks, 18–21, 43–52, 99, 100, 103
START I/O, *see* Input/Output
State, privileged, *see* Control, states
Storage-storage architecture, 58–60, 157
SOLOMON, 231
Subroutine, 94, 97
Subroutine linkage, 96–105, 153
Subroutine linkage instructions, 77, 78
Superconductor, 292
Swapping in virtual memory, 123–124
Switching speed, 180

Symbol tables, 18–20, 24
Symmetric instruction sets, 79

Table lookaside buffer, 128–129, 264
Tags, 36, 167
Task switch, 42
Technological advances, 290
TEST I/O, *see* Input/Output
Three address forms, 61
Time-sharing, 9
Top-of-stack pointer, 43–46
Transfer, *see* Branch
Traps, 107
Two address forms, 61
Two's complement, 29

UNIVAC (I), 96, 137, 290
UNIVAC 1100, 31, 42, 53, 65, 69, 81–82

VAX-11, 54, 66, 79–80, 83, 91, 92, 103, 116, 117
Vector operations, 231, 232

Words, concept, 31–32, 34
Working set, 124

Zone portions, 30